The Physics

V. C. REDDISH

# of Stellar Interiors

The Physics of Stellar Interiors. An Introduction

V. C. REDDISH

Edinburgh University Press

Edinburgh University Press
22 George Square, Edinburgh
ISBN 0 85224 270 0
Printed in Great Britain by
Aberdeen University Press

PREFACE

This book is essentially the text of a course of fifteen lectures given to astronomy students at the Warner and Swasey Observatory, Case-Western Reserve University, Cleveland, Ohio, in the Spring of 1969.

There are many excellent textbooks on stellar structure but they were either too descriptive or too advanced for my purposes. I hope that this book fills a gap. It is designed to serve the needs of honours undergraduates in astronomy and in physics, or as an introduction to the subject for graduate students and teachers.

The treatment of the subject is intended to give the reader an understanding of the physical conditions and processes inside stars, without requiring more knowledge of physics and mathematics than might be expected of an average student after two years in a University. Approximations and simplifications have been used to this end, so that although the treatment is mathematical rather than descriptive, it is fairly simple and the reader does not have to be much of a mathematician to understand it.

Teaching is also a learning process, and the lectures on which this book is based benefited from the lively interest and searching questions of a score of graduate students at the Warner and Swasey Observatory. The opportunity to give the lectures was provided by the Director of that time, Dr S.W. McCuskey, to whom the author is indebted for that and many other kindnesses.

V.C.R. *Edinburgh, May 1973*

CONTENTS

CONTENTS

CHAPTER ONE

I. Introduction.

A study of stars and galaxies enables us to examine the behaviour of matter in conditions which cannot be reproduced in the laboratory. Astrophysics may be regarded as the physics of extreme environments.

In the centres of stars, temperatures are commonly some 10^7 K, but in late stages of stellar evolution they exceed 10^9 K. At these temperatures, the nuclei of atoms break down. Densities are generally similar to those of solid materials on Earth, although the temperatures are so high that the matter is gaseous; but in dying stars densities increase by factors that range from 10^6 to an incredible 10^{15} or more, and matter takes on strange properties. In contrast, between the galaxies, one atom per cubic metre may be all that can be expected.

However, it is not just the extreme values of temperature and density which make the physical conditions of such interest, but the enormous values of mass, distance, time and velocity within which they operate. It is possible to heat a small quantity of matter to above 10^6 K in the laboratory for a short period of time; but to heat 10^{30} kg to that temperature for 10^{10} yr, or to gather 10^{40} Kg together in one massive object, or to move such agglomerations with speeds approaching that of light for similarly long periods of time; these are situations which only the stars and galaxies can present us with, and only by studying them can we hope to discover how matter behaves in such circumstances.

It may therefore seem all the more surprising that so much of what is observed can be accounted for in terms of simple physical laws that are based on experience of events on Earth, but this serves to show the universality of these laws. Newton's laws of gravity and of motion, the first two laws of thermodynamics with Einstein's law of the equivalence of mass and energy, Boyle's law and Charles' law of perfect gases, and Heisenberg's Uncertainty Principle, form the basis of most of the contents of this book.

One of our most common experiences is that heat always flows from hotter to colder regions, an experience formalised as the Second Law of Thermodynamics. Since heat is flowing out of the Sun it follows that its interior must be hotter than its visible surface; just how much hotter is a matter for calculation in a subsequent chapter.

Geological evidence indicates that the Earth is some 4 500 millions of years old and that the Sun has been pouring out heat for a comparable time. This information is important to our understanding of the Sun and other similar stars in two ways. Firstly, it shows that the Sun is stable over such long periods of time, and secondly, that it draws on a vast supply of thermal energy.

The stability requires that the gravitational self-attraction, tending to make the Sun collapse in on itself, is balanced by the expansive pressure of the hot gas of which it is composed, and this balance must be self-restoring if it is slightly disturbed; the stability requirements and how they are met in practice will be examined through elementary thermodynamics.

The conservation of energy, embodied in the First Law of Thermodynamics, will lead us to conclude that we cannot account for the vast expenditure of energy by the Sun and stars unless we turn to Einstein's Law $E = Mc^2$; transmutation of the elements, hydrogen to helium, helium to heavier elements, provides most of the energy requirements for most stars; but not all the energy, because gravitation provides some, and particularly when we come to study the extraordinary properties of supermassive objects we shall find that it sometimes dominates all else.

Energy production by nuclear synthesis is very sensitive to temperature; a very small rise in temperature causes an enormous increase in heat produced, and again raises the question as to how stability is maintained. It turns out that when more heat is added to the gas than can escape, causing the gas to expand, the work done in expansion exceeds the thermal energy added. Consequently the gas cools, and the amount of energy produced by nuclear reactions falls, until equilibrium is regained.

Similarly underproduction of energy causes contraction, a rise in temperature, and an increase in energy produced. Thus, negative feedback ensures the stability of the star.

This stability cannot last indefinitely because the nuclear energy supply, although enormous, is nonetheless limited. Higher temperatures are required to produce heavier elements, and so as nuclear synthesis proceeds the star contracts stage by stage, becoming more dense as well as hotter. As the density rises, each atom has less and less space available – that is to say, the position of each atom becomes more and more accurately defined. However, Pauli's Exclusion Principle sets a limit to the accuracy with which the position can be determined at a given momentum; at this limit, if the error in position is reduced, the momentum must be increased. This happens in the dense cores of the stars, with the result that the momenta of the atoms becomes determined by the density and not by the temperature; in this condition the material is said to be *degenerate*. Now it follows from Newton's Second Law of Motion that pressure (force per unit area) is rate of change of momentum, and consequently the pressure in degenerate stars is determined by density and not by temperature. The negative feedback which had ensured stability depended on the relationship between pressure and temperature and no longer applies; indeed, nuclear synthesis in degenerate material is potentially explosive and in this situation we shall find the cause of some of those stellar explosions, the supernovae.

Degenerate material can become so dense that all the protons and electrons are not permitted by the Exclusion Principle to exist as separate entities, and many are forced to fuse together to become neutrons. These neutron stars possess the most extraordinary properties, the atoms forming a material vastly stiffer than steel, through which the neutrons flow as easily as an unobstructed superfluid.

At least as strange and exciting as neutron stars are the supermassive objects and black holes. Once again we have recourse to Einstein's Law $E=Mc^2$, this time to take account of the gravitational attraction of the mass equivalent of the energy. In supermassive objects this becomes very large and produces rather startling behaviour. Whereas stars of normal mass lose energy as they contract in semi-stable equilibrium, radiating away part of the released gravitational energy and storing the rest as heat, supermassive objects require additional energy to maintain equilibrium as they contract. The violently unstable situation that results may produce explosions on the scale of whole galaxies, and perhaps are what we observe as quasars.

Stable stars, nuclear synthesis and energy generation, red giants and white dwarfs, neutron stars and supermassive objects; all are studied in this book by simple extensions of quite elementary laboratory physics that account satisfactorily for the extraordinary phenomena observed. Astrophysics is here regarded as a natural extension of earthbound physics to the study of the behaviour of matter in extreme physical conditions, and it is reasonable to suggest that such a study is essential to a proper understanding of the nature of the physical world.

Stable Stars.

PART ONE

Structure, Energy Generation and Nuclear Synthesis

2.1. *A Summary of Data.*

Measurements of the intensity of radiation from a star as a function of wavelength give information about the surface layers of the star emitting that radiation. In particular they enable us to determine the effective temperature, pressure and chemical composition of the gas in the surface layers and to estimate the total amount of energy being radiated – the so-called luminosity. Similarly, measurements of the motions of a pair of stars in orbit about their common centre of gravity enable the masses of the components to be calculated. The Sun is found to have the following properties:*

$$\left.\begin{array}{l}\text{mass: } M=2\times 10^{33}\,\text{g}\\ \text{luminosity: } L=3.9\times 10^{33}\,\text{erg s}^{-1}\\ \text{surface effective temperature: } T_e=5800\text{ K}\\ \text{radius: } R=7\times 10^{10}\,\text{cm}\end{array}\right\}\quad(1)$$

In this way a fund of physical data about the stars has been collected and it shows a number of systematic features. For example, when the luminosities of stars are plotted against their surface effective temperatures, most of the points fall along a line

$$\log L \approx 7.5 \log T_e + 5.44 \qquad (2)$$

and this has come to be known as the *main sequence*. It is shown on figure 1. Now the number of points in any small range of L, T_e on that figure must be proportional to the number of points passing through that range and to the time they spend within it. Consequently we deduce that most stars either are formed on, or spend most of their time on, the main sequence.

* The centimetre gram second (c.g.s.) system is in common use in astronomy and has been retained throughout this book since most other texts and papers to which the student may turn in due course employ this system of units.

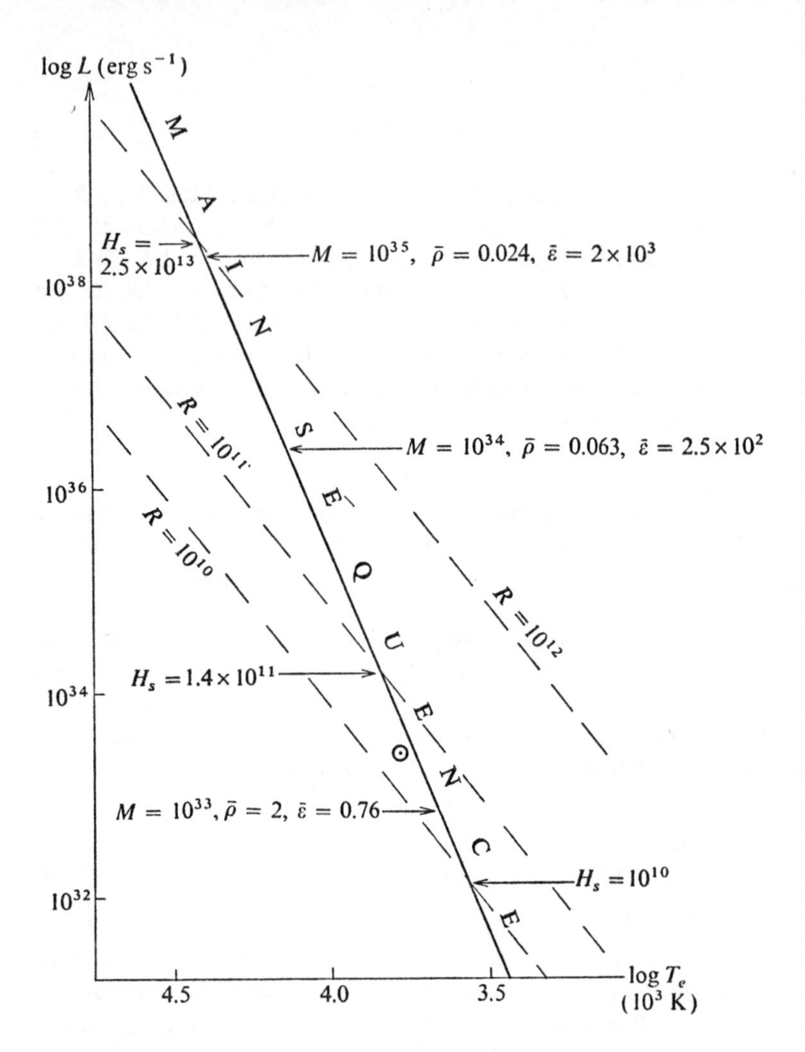

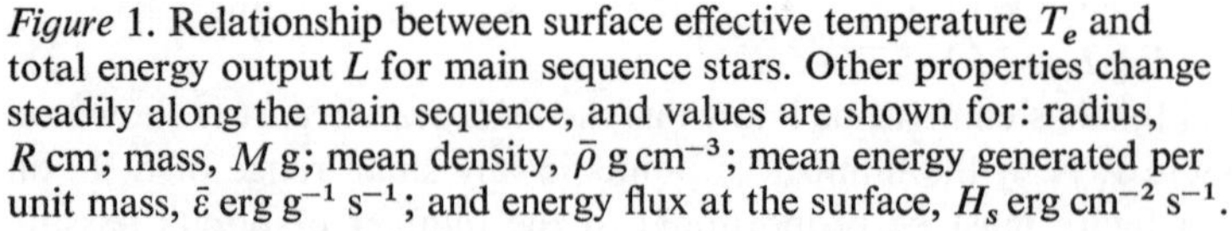

Figure 1. Relationship between surface effective temperature T_e and total energy output L for main sequence stars. Other properties change steadily along the main sequence, and values are shown for: radius, R cm; mass, M g; mean density, $\bar{\rho}$ g cm^{-3}; mean energy generated per unit mass, $\bar{\varepsilon}$ erg g^{-1} s^{-1}; and energy flux at the surface, H_s erg cm^{-2} s^{-1}.

Lines of constant radius can be put on figure 1, since by Stefan's Law $L = 4\pi R^2 \sigma T_e^4$, or

$$\log L = 2 \log R + 4 \log T_e - 3.15 \qquad (3)$$

The masses of main sequence stars are also found to be uniquely related to their luminosities and the relationship is known as the mass-luminosity law:

$$\left.\begin{array}{ll}\log L = 3.85 \log M - 94.2 & \text{for } M \leqslant 7.8 \times 10^{33} \\ \log L = 1.85 \log M - 26.6 & \text{for } M > 7.8 \times 10^{33}\end{array}\right\} \quad (4)$$

Appropriate values of M are noted on figure 1.

The radii of main sequence stars can be calculated by eliminating T_e between (2) and (3) giving

$$\log L = 4.29 \log R - 13.0 \quad (5)$$

Similarly, relationships can be obtained for the thermal flux $H_s = L/4\pi R^2$; for the mean density $\bar{\rho} = 3M/4\pi R^3$, and for the rate of energy generation per unit mass, $\varepsilon = L/M$. It may be noted that M, R, H_s and ε all increase towards higher luminosities (figure 1), but $\bar{\rho}$ decreases. It will be of later interest to note more explicitly the dependence of radius on mass:

$$\left.\begin{array}{ll}R \propto M^{0.9} & \text{for } M \leqslant 7.8 \times 10^{33} \\ R \propto M^{0.43} & \text{for } M > 7.8 \times 10^{33}\end{array}\right\} \quad (6)$$

We can now consider what we have to account for among the properties of main sequence stars. As shown above and illustrated on figure 1, it is that a main sequence star of a given mass has unique values of luminosity and radius, and hence of surface temperature and density. We seek to discover why this should be.

2.2. *A Brief Introduction to the Theory.*

We shall begin with a brief examination of some of the theoretical ideas, using simple and approximate mathematical relationships to illustrate them and to bring to notice the physical problems and relationships which we shall examine in more detail later. At this stage we will also assume some things which will be derived or justified subsequently, but this preliminary skirmish with the theory will help us to get the feel of the physical situation before delving more deeply into detail.

a) *Equilibrium.* Suppose that a cloud of gas has no internal pressure; it is easy to show from the Law of Gravity and Newton's Third Law of Motion that it will collapse, due to gravitational self-attraction, at free-fall rate in a time

$$t \sim (6\pi G\rho)^{-\frac{1}{2}} \quad (7)$$

For example, consider the Sun, which has a mean density $\bar{\rho} = 1.4 \text{ g cm}^{-3}$; it would collapse in a time 10^3 s. But the age of the Earth is $\geqslant 4.5 \times 10^9$ yr, and the Sun must be at least as old. Consequently the Sun must have an internal pressure great enough to balance, at least approximately, the force of gravitational self-attraction. And this must be the case for other stars – it would be unreasonable to suppose that the Sun is exceptional in this respect, and in any case some individual stars are known from ancient records to have been in existence for many hundreds of years, $\sim 10^{10}$ s.

The internal pressure needed to balance the gravitational attraction is

$$P = (\text{gravitational force/unit mass}) \times (\text{mass/unit area})$$

$$\propto \left(\frac{GM}{R^2}\right) \cdot \left(\frac{M}{R^2}\right)$$

that is,

$$P \propto \frac{GM^2}{R^4} \tag{8}$$

where the constant of proportionality is determined by the radial distribution of mass in the star, and the particular radial distance (the fraction of R) at which P is measured.

In a perfect gas having molecular weight μ,

$$P = \frac{\rho \mathcal{R} T}{\mu} \tag{9}$$

where $\mathcal{R}$ is the gas constant. Also, the density

$$\rho \propto \frac{M}{R^3} \tag{10}$$

where again the constant of proportionality depends on the radial mass distribution and the radial distance. Substituting for ρ from (10) into (9), and dividing the result by (8) gives

$$T \propto \frac{\mu M}{R} \tag{11}$$

Once more the constant of proportionality depends on the mass distribution and the radial position. However, in the particular case at the centre of the star,

$$T_c \propto \frac{\mu M}{R} \tag{12}$$

where the constant of proportionality depends only on the radial distribution of mass.

If we refer back to (6), which shows how the radii of main sequence stars are related to their masses, then it follows from (11) that if main sequence stars are of similar structure (that is, if they have the same *relative* distributions of mass with radius), then their central temperatures depend on their masses as follows:

$$\left.\begin{aligned} T_c &\propto \mu M^{0.1} \quad \text{for } M \leqslant 7.8 \times 10^{33} \text{ g} \\ T_c &\propto \mu M^{0.57} \quad \text{for } M > 7.8 \times 10^{33} \text{ g} \end{aligned}\right\} \tag{13}$$

Therefore T_c is approximately constant for main sequence stars with masses $M \leqslant 4M_\odot$, and increases slowly for more massive stars (unless the composition and/or the structure change with mass).

For equilibrium, energy produced = energy radiated. We shall see later that the rate of energy production is very highly temperature sensitive. Thus very small changes in central temperature are adequate to balance large differences in luminosity, and this suggests the reason for the near constancy of central temperature.

b) *Energy Flow.* We can state as a general rule, flow $\propto$ pressure gradient/resistance (compare Ohm's Law $I = V/R$). Thus, radiant energy flow per unit area $\propto$ radiation pressure gradient/opacity per unit volume. That is,

$$H \propto \frac{d(\frac{1}{3}aT^4)/dr}{\kappa\rho} \propto \frac{T^3\,dT/dr}{\kappa\rho} \tag{14}$$

where κ denotes opacity per unit mass at temperature T and density ρ, and a is the radiation constant. If the whole of the energy flowing out through the star is transported by radiation,

$$L \propto 4\pi R^2 H \propto \frac{R^2T^3\,dT/dr}{\kappa\rho} \tag{15}$$

and since *for a given structure* $T \propto T_c$, $dT/dr \propto T_c/R$, and $T_c \propto \mu M/R$, it follows that (15) becomes

$$L \propto \frac{\mu^4 M^3}{\kappa} \tag{16}$$

Also, the opacity per unit mass, κ, is a function of density, temperature and chemical composition; that is, $\kappa = \kappa\ (\rho, T, \text{composition})$. For the solar composition, Kramer derived a power law approximation for the opacity (which we shall study in greater detail later),

$$\kappa \propto \rho T^{-3.5} \tag{17}$$

This converts, using (10) and (11), to

$$\kappa \propto \mu^{-3.5} M^{-2.5} R^{0.5} \tag{18}$$

Consequently (16) and (18) give

$$L \propto \mu^{7.5} M^{5.5} R^{-0.5} \tag{19}$$

At lower density or higher temperature, the opacity is due mainly to electron scattering, for which κ = constant, the value of the constant being determined by the chemical composition. Using this, (16) becomes

$$L \propto \mu^4 M^3 \tag{20}$$

The observational data represented by (6) and (11) has already shown us that with increasing stellar mass, main sequence densities decrease rapidly while temperatures increase slowly. Consequently we would expect low-mass stars to have a mass-luminosity law as in the proportionality (19) and higher masses to tend towards the relationship (20). The data is represented by equations (4), which show for low-mass stars $L \propto M^{3.85}$, and for higher masses $L \propto M^{1.85}$. Our predicted indices are rather too high, but change by the correct amount as the masses change. The predicted indices would change if the opacity law changes, and this may be considered to be a possible source of error; also the assumption that the energy is transferred throughout the star by radiation may be incorrect – we have ignored the possible effect of heat transport by convection.

We can now account in principle for the property of the main sequence which was noted in the last paragraph of section 2.1. For a given T_c, M determines R through (11) and L through (19) or (20); thus L and R, and hence L and T_e, are determined primarily by M with T_c nearly constant; the main sequence is the locus of values of (L, T_e) for stars of various masses.

The following points are worth noting:

1. The numerical value of κ, the absorption coefficient per unit mass, is of order unity.

2. Energy fluxes at the surfaces of main sequence stars are in the range 10^{10} to 10^{14} erg cm^{-2} s^{-1}; the energy flux in the interior is higher because the same energy is crossing a smaller area.

3. The energy content, Q, of a gas is given by

$$Q = \int dQ = \int C_V dT + \int P dV = \int C_P dT \text{ for a perfect gas,}$$
$$= \tfrac{5}{2}\mathcal{R}T = 2 \times 10^8 T \text{ erg mol}^{-1} \qquad (21)$$

For example, at the centre of the Sun, $\mu = \frac{1}{2}$, $\rho = 100$, $T = 10^7$K, and hence $Q = 4 \times 10^{17}$ erg cm^{-3}. The average value throughout the Sun's volume of 1.4×10^{33} cm^3 is 10^{15} erg cm^{-3}, giving a total energy content Q(total, Sun) $= 1.4 \times 10^{48}$ erg. Since the Sun is losing heat at a rate $L = 4 \times 10^{33}$ erg s^{-1}, the cooling time scale is

$$t(\text{cooling}) \sim \frac{Q(\text{total})}{L} = 3 \times 10^{14} \text{ s} = 10^7 \text{ yr}$$

This long cooling time resulting from large thermal capacity is a major factor in ensuring the Sun's stability.

The radiant energy content is

$$Q(\text{rad}) = aT^4 = 7.5 \times 10^{-15} T^4 \text{ erg cm}^{-3} \qquad (22)$$

At the centre of the Sun this is 7.5×10^{13} erg cm^{-3}. The average throughout the volume of the Sun is 2×10^{13} erg cm^{-3}, so that the total radiant energy content of the Sun is

$$Q(\text{rad, total}) = 3 \times 10^{46} \text{ erg},$$

one-fiftieth of the total energy content.

The time for radiation to travel from the centre to the surface is:

$$t(\text{rad}) \sim \frac{Q(\text{rad, total})}{L} = 7\times10^{12}\ \text{s} = 5\times10^{5}\ \text{yr}$$

(since in equilibrium the source of the radiant energy must supply L; if all the radiant energy of the Sun were removed, a source at the centre would require a time t(rad) to replenish it, to the surface).

Compare this time to the time for radiation to travel a distance equal to the solar radius of 7×10^{10} cm in free space, 2.3 seconds. The difference shows the enormous effect which the opacity of the gas has, by absorption, re-emission and scattering, in slowing down the passage of radiation through it.

4. The velocity of sound, the speed of a pressure wave through gas, is

$$v = \left(\frac{\gamma P}{\rho}\right)^{\frac{1}{2}}$$

For a perfect gas, $P/\rho = \mathscr{R}T/\mu$ and hence

$$v = \left(\frac{\gamma \mathscr{R} T}{\mu}\right)^{\frac{1}{2}} \tag{23}$$

At the centre of the Sun, $\mu = \frac{1}{2}$ and $T = 10^7$ K, and since $\gamma = 5/3$ and $\mathscr{R} = 8.3\times10^7$,

$$v(\text{centre of Sun}) = 5\times10^7\ \text{cm s}^{-1}.$$

Since the velocity varies as $T^{\frac{1}{2}}$, it is not much less than this on average through the Sun; that is, $\bar{v}(\text{Sun}) \sim 10^7$ cm s^{-1}, and consequently the time taken by a pressure wave to travel from the centre to the surface and back to the centre, which is the natural period of pulsation of the Sun, will be of the order of a few hours.

5. Suppose that the heat of the Sun and stars is carried outwards by convection, rather than by radiation as we have supposed. Assume that at any distance from the centre, half the material is flowing outwards with a temperature T and some speed v, and half is flowing inwards with temperature $T-\Delta T$ and speed v. Then the heat transported is

$$H = \frac{Q\rho v}{\mu}\ \text{erg cm}^{-2}\ \text{s}^{-1} = 2\times10^{8}\frac{\rho v \Delta T}{\mu} \tag{24}$$

using (21).

The mean density of the Sun is $\rho = 1.4$ g cm^{-3}, the mean molecular weight is $\mu = \frac{1}{2}$, and the heat transport at the surface is $H_s = L/4\pi R^2 = 6\times10^{10}$ erg cm^{-2} s^{-1}. Inserting these in (24) gives:

$$v\Delta T = 10^2\ \text{K cm s}^{-1}$$

If we compare this to the energy which would be transported at the speed of sound (a pressure wave) at a position of average temperature in the Sun, where $\bar{T} \sim 10^6$ K,

$$\bar{v}\bar{T} \sim 10^6 \times 1.6 \times 10^7 = 1.6 \times 10^{13} \text{ K cm s}^{-1}$$

we can see that very slight convection could easily transport the solar output of heat. The question as to whether or not the interior of stars are convective is therefore a matter of considerable importance. Consequently it is of interest to consider the conditions necessary for convection to take place, but before doing so it is useful to determine the relationship between temperature and pressure in a star in convective equilibrium.

c) *Convective Equilibrium.* Consider volumes of gas moved adiabatically from one region to another, then giving up to or receiving energy from their surroundings. During the adiabatic change no heat is lost or gained:

$$dQ = C_V dT + P dV = 0$$

and since $PV = \mathcal{R}T$ we can write

$$\frac{C_V dT}{\mathcal{R}T} + \frac{P dV}{PV} = 0 \quad \text{or} \quad \frac{C_V}{\mathcal{R}} d \log T + d \log V = 0$$

Integrating,

$$VT^{C_V/\mathcal{R}} = \text{constant},$$

and since $V = \mu/\rho$ this gives

$$\rho \propto T^{C_V/\mathcal{R}}$$

But $\mathcal{R} = C_P - C_V$ and hence $C_V/\mathcal{R} = 1/(\gamma - 1)$ where $\gamma = C_P/C_V$. Consequently

$$\rho \propto T^{1/(\gamma-1)} \tag{25}$$

and since $P \propto \rho T$ it follows that

$$P \propto T^{\gamma/(\gamma-1)} \tag{26}$$

This, then, gives the relationship between the pressure and temperature distributions within a star in convective equilibrium.

d) *Condition for Convection.* The density, temperature and pressure are all functions of radius within the star; that is $\rho(r)$, $T(r)$ and $P(r)$. Consequently we can write $P(T)$. Let us represent this dependence *locally* by a power law approximation, $P \propto T^n$. Since ρ, T and P are all decreasing functions of r, n must be positive.

Consider an element moved adiabatically from a region (P_1, T_1) outwards to a region (P_2, T_2), where $P_2 < P_1$. The element will change density and temperature as necessary to maintain pressure equilibrium with its surroundings, its temperature becoming T_3, say. Thus:

Surroundings (region 1) P_1, T_1: (region 2) $P_2, T_2 = T_1(P_2/P_1)^{1/n}$
Element (region 1) P_1, T_1: (region 2) $P_2, T_3 = T_1(P_2/P_1)^{(\gamma-1)/\gamma}$

where γ is the ratio of specific heats at constant pressure and constant volume (see (c) above).

Now $P_2 < P_1$. Therefore if $(\gamma-1)/\gamma < 1/n$, then $T_3 > T_2$ and the element gives up heat to its surroundings. This heat is transported down the temperature gradient, that is outwards, by convection. Rising elements, being hotter than their surroundings, are lighter and tend to rise further; so the convection is driven by the thermal energy.

Conversely, if $(\gamma-1)/\gamma > 1/n$, then $T_3 < T_2$ and the element gains heat from its surroundings; whereas elements moving inwards to higher (P, T) give up heat to their surroundings, transporting heat against the temperature gradient. In this case rising elements are cooler than their surroundings and tend to fall unless driven by some energy source other than the thermal energy.

Consequently the condition for convection is

$$\frac{(\gamma-1)}{\gamma} < \frac{1}{n}$$

Calculations of $\rho(r)$, $T(r)$, and $P(r)$ for given opacity and energy sources show generally $n \approx 4$; the condition for convection is therefore that $\gamma < 4/3$. But γ(radiation) $= 4/3$ and γ(perfect gas) $= 5/3$: therefore for a mixture, $\gamma > 4/3$. It follows that there is no convection and the earlier assumption that the heat is transferred by radiation is justified.

The condition for convection may be written, using (26),

$$\frac{d \log T}{d \log P} = \frac{1}{n} > \frac{\gamma-1}{\gamma} \qquad (27)$$

If the energy sources are concentrated towards the centre, so that the centre heats up excessively, or if the opacity is so high that a steep temperature gradient builds up to transport the heat, then (27) may be satisfied locally. This does occur: the high temperature-dependence of the rate of nuclear energy production causes it to be very centrally concentrated, producing a convective core in the star; and the opacity in the hydrogen ionisation zone (regions where $T \simeq 10^4$ K) is so high that it leads to an outer convective zone, close to the surface of the star. These convection zones change the interior structure of stars, and our constants of proportionality which depend on them.

e) *Radiative Equilibrium.* The total pressure P at any point in the star is the sum of the gas pressure and the radiation pressure, thus

$$P = P_g + P_r$$

where $P_g = \rho \mathscr{R} T/\mu$ and $P_r = \frac{1}{3} a T^4$. Writing $P_g = \beta P$, and hence $P_r = (1-\beta)P$, it follows that

$$P = \frac{\rho \mathscr{R} T}{\mu \beta} = \frac{a T^4}{3(1-\beta)}$$

from which

$$\rho = \left\{ \frac{a \mu \beta}{3 \mathscr{R} (1-\beta)} \right\} T^3 \qquad (28)$$

If β is constant (and this is approximately true) this is of the form

$$\rho = K T^n \qquad (29)$$

which is the equation of a polytropic gas sphere. Thus we can say that if the ratio of gas pressure to radiation pressure is constant in a star, it will be a polytrope of index $n = 3$ whereas a convective region (equation 25 with $\gamma = 5/3$) has index $n = 3/2$. This is particularly useful because the properties of polytropes, such as the dependence of temperature and density on radial distance, are well known and tabulated.

Summarising briefly: Observations show that for most stars, the so-called main sequence stars, unique values of luminosity L and radius R are associated with any given mass M. They further show that M/R has mainly the same value for all these stars, and that $L \propto M^x$ where $1.85 \leqslant x \leqslant 3.85$.

Our preliminary theoretical considerations have shown that the central temperature $T_c \propto \mu M/R$, the constant of proportionality being determined by the interior structure. Comparison with observational data indicates that main sequence stars have a common structure and closely similar central temperatures.

We have also shown that $L \propto M^3/\kappa$ where κ is the opacity per unit mass and generally varies as $\kappa \propto M^y$ where $0 \leqslant y \leqslant 2$; consequently $L \propto M^z$ where $3 \leqslant z \leqslant 5$. The index in this theoretical mass-luminosity relation being higher than observed, we examined our assumptions of radiation equilibrium to determine whether or not it is correct or if perhaps convection is transporting the heat.

The requirement for convection to occur turned out to be a very steep gradient of temperature relative to pressure,

$$d \log T / d \log P > (\gamma - 1)/\gamma$$

where $\gamma = C_P/C_V$ and takes the value 4/3 for radiation and 5/3 for a perfect gas. The temperature gradients in stars are generally less than this except near the centre and in the hydrogen ionisation zone close to the surface. Thus a star in radiative equilibrium is approximately a polytrope of index 3, with a core having index 3/2.

In passing, we noted the large thermal inertia of stars, the heat content of the Sun being so great that its cooling time is 10^7 yr; we noted also that the Sun's natural period of vibration is only a few hours, and that if all the interior pressure were removed it would collapse in 10^3 s; consequently the large thermal inertia is a major factor in ensuring the stability of main sequence stars.

2.3. *The Equations of Stellar Structure.*

Now let us consider models of main sequence stars in more detail. We assume a spherically symmetric cloud of gas in hydrostatic equilibrium, with energy transport by radiation and no magnetic fields or rotation.

1. *Equation of Distribution of Mass.* If M_r represents the mass contained within the radius r, and ρ is the density at r, then

$$M_r = \int_0^r 4\pi\rho r^2 \, dr \qquad (30)$$

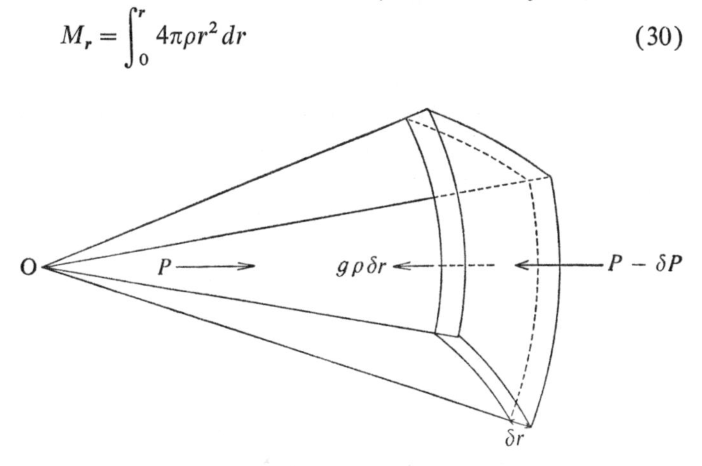

Figure 2. The forces of pressure and gravity acting on unit area of a thin shell of matter at any distance r from the centre of a star.

2. *Equation of Hydrostatic Equilibrium.* Consider the forces acting on unit area of a shell of material of thickness δr at r from the centre O, figure 2. For equilibrium,

$$\delta P = -g\rho\,\delta r$$

where $g = GM_r/r^2$ is the gravitational force per unit mass at r due to the attraction of the mass interior to r. Thus,

$$\frac{dP}{dr} = -\frac{GM_r\rho}{r^2} \qquad (31)$$

3. *Equation of Radiative Energy Transport.* Let the flow of the energy of radiation across unit area in unit time be represented by F, and let κ denote the mass absorption coefficient; the fraction of radiation absorbed in traversing a mass $\rho\,\delta r$ (figure 2 again) is then

$$\frac{\delta F}{F} = \kappa\rho\,\delta r \tag{32}$$

The momentum transferred from the radiation to this material must be the change in radiation pressure over δr (since pressure is simply transfer of momentum); thus

$$\delta P(\mathrm{rad}) = -\frac{\delta F}{c} = -\frac{F\kappa\rho\,\delta r}{c}$$

Consequently,

$$F = -\frac{c}{\kappa\rho}\frac{dP(\mathrm{rad})}{dr} \tag{33}$$

If we are outside energy producing regions and the star is in equilibrium, then the total flow of radiation across the whole spherical surface at r will be constant and equal to the luminosity L. So, since $P(\mathrm{rad}) = \frac{1}{3}aT^4$ we have

$$L = 4\pi r^2 F = -\left(\frac{4\pi r^2 c}{\kappa\rho}\right)\frac{d}{dr}\left(\frac{aT^4}{3}\right)$$

or

$$L = -\left(\frac{16\pi ac}{3\kappa\rho}\right)r^2 T^3 \frac{dT}{dr} \tag{34}$$

4. *Equation of Energy Production.* If $L(r)$ denotes the outflow of energy across the spherical surface of radius r, so that for instance $L = L(R)$, then in equilibrium the average energy production per unit mass at the radius r is

$$\varepsilon = \frac{dL(r)}{dM_r} \tag{35}$$

This will be a particularly useful formula when we deal with nuclear energy production, for which ε will depend on $T(r)$, $\rho(r)$ and the chemical composition.

We may re-write these four equations with M_r as the independent variable, to give from (30),

$$\frac{dr}{dM_r} = \frac{1}{4\pi r^2 \rho} \tag{36}$$

from (30) and (31):

$$\frac{dP}{dM_r} = -\frac{GM_r}{4\pi r^4} \tag{37}$$

from (30) and (34):

$$\frac{dT}{dM_r} = -\frac{3\kappa L}{64\pi^2 acr^4 T^3} \tag{38}$$

and from (35):

$$\frac{dL}{dM_r} = \varepsilon \tag{39}$$

Equations (36) to (39) are the four basic equations expressing the structure of spherically symmetric stars in quasi-static equilibrium, representing respectively distribution of mass, hydrostatic equilibrium, energy transport and energy generation.

Three variables in the equations are functions which require further specification; they are:

the equation of state: $P = P(\rho, T, \text{composition})$ (40)
the opacity: $\kappa = \kappa(\rho, T, \text{composition})$ (41)
and the energy source: $\varepsilon = \varepsilon(\rho, T, \text{composition})$ (42)

and these require that the composition be specified as a function of M_r.

We now have four dependent variables, $r(M_r)$, $\rho(M_r)$, $T(M_r)$ and $L(M_r)$ to be calculated from four differential equations, and we require four boundary conditions; these are:

$$\left.\begin{aligned} &r = 0, L = 0 \text{ at } M_r = 0 \\ &\rho = 0, T = 0 \text{ at } M_r = M(R) \end{aligned}\right\} \tag{43}$$

The structure of the star can now be calculated. Note that the four boundary conditions ensure that this can be done, but they do not ensure that there is a single unique solution.

If the energy is produced by nuclear reactions, then nuclear synthesis will cause the composition to change and the calculated structure will refer to one particular epoch only. We have

$$\begin{aligned} \text{composition} &= f(\text{structure, time}) \\ &= f(\rho, T, \text{composition}, t) \end{aligned}$$

so that

$$\frac{\partial}{\partial t}(\text{composition}) = f(\rho, T, \text{composition}, t) \tag{44}$$

Equations (36) and (39) should then be written as partial derivatives. If there is convection, this will transport energy and matter and hence require modification to (38) and to (44).

We have still to consider in more detail the equation of state, the opacity and the energy generation, but before doing so let us determine the conditions which must be met for the stability which has been assumed.

2.4. *Stability*.

For a star to be stable, the internal expansive forces – the sum of the various pressures and centrifugal force – must be equal to the contractive forces – the sum of external pressures and self-gravitation. The *Virial Theorem* is a statement of these forces.

For simplicity let us suppose that a star is composed of atomic particles all of one kind, each particle having mass m and position $\bar{r}$ and being subject to a force $\bar{F}$. Then the acceleration of the moment of inertia of the assembly of particles resulting from the net forces acting upon it is

$$\frac{d^2I}{dt^2} = \frac{d^2}{dt^2}\sum(m\bar{r}^2) = \sum 2\left[m\left(\frac{dr}{dt}\right)^2 + \bar{r}\bar{F}\right]$$

That is,

$$\tfrac{1}{2}\frac{d^2I}{dt^2} = 2(\text{kinetic energy}) + \sum \bar{r}\bar{F} \tag{45}$$

The term $\sum \bar{r}\bar{F}$, which is a measure of the forces of interactions between the particles, is *Clausius' Virial*, and was derived by him in this form in an attempt to account for forces of interaction between molecules in a gas, forces which were also accounted for in a different manner by van der Waals.

The *Virial* is analogous to the work done by the forces $\bar{F}$ in moving the particles to $\bar{r}$ and therefore represents the potential energy in the system; so we can write:

$$\tfrac{1}{2}\frac{d^2I}{dt^2} = 2(\text{kinetic energy}) + \text{potential energy} \tag{46}$$

For stability, the *net* forces acting to change the distribution of particles as a whole must be zero; consequently there will be no acceleration of the moment of inertia and the condition for stability may therefore be written

$$2(\text{kinetic energy}) + \text{potential energy} = 0 \tag{47}$$

Particles may be redistributed locally within the assembly (such as may result from convection) without net forces necessarily acting on the system as a whole and so without necessarily introducing instability.

The potential energy of the star is the gravitational potential energy Ω; thus

$$\text{potential energy} = \Omega \tag{48}$$

The kinetic energy is the thermal energy, thus

$$\begin{aligned}\text{kinetic energy} = \text{thermal energy} &= \tfrac{3}{2}\mathcal{R}T = \tfrac{3}{2}(C_P - C_V)T \\ &= \tfrac{3}{2}(\gamma-1)C_VT \\ &= \tfrac{3}{2}(\gamma-1)U\,\text{mol}^{-1}\end{aligned} \tag{49}$$

where U is the internal heat given by $dU = C_V dT$ mol^{-1}. Thus the requirement for stability is given by (47), (48) and (49) as

$$3(\gamma - 1)U + \Omega = 0 \tag{50}$$

The total energy E is $E = U + \Omega$, or:

$$E = \frac{3\gamma - 4}{3(\gamma - 1)}\Omega \tag{51}$$

using (50). For stability the total energy must be negative (if it is positive the system will disperse); it follows from (51) that for stability

$$\gamma > 4/3 \tag{52}$$

To determine the value of γ for a star, consider first the value for a perfect gas. The total internal energy U of a perfect gas is due to the kinetic energy of its particles, that is,

$$U = \tfrac{3}{2}NkT = \tfrac{3}{2}\mathscr{R}T \text{ mol}^{-1}$$

where N is Avogadro's number, k is Boltzmann's constant and $\mathscr{R}$ the gas constant. Each particle has three degrees of freedom, three orthogonal independent directions of motion carrying energy. Since by the principle of equipartition of energy the energy must be divided equally between the degrees of freedom, the energy per degree of freedom per mole is $\tfrac{1}{2}\mathscr{R}T$. Consequently, if there are n degrees of freedom;

$$U = \frac{n}{2}\mathscr{R}T \tag{53}$$

But

$$C_V = \left(\frac{\partial U}{\partial T}\right)_V = \frac{n}{2}\mathscr{R} \tag{54}$$

and since $\mathscr{R} = C_P - C_V$ it follows that

$$\frac{C_P}{C_V} = \gamma = 1 + \frac{2}{n} \tag{55}$$

For a perfect gas $n = 3$ and so $\gamma = 5/3$.

Radiation has six degrees of freedom; two orthogonal independent directions of vibration (polarisation) of the electromagnetic wave perpendicular to each of the three orthogonal independent directions of propagation. (Since a gas has no stress as in a field or in a solid, it can only transmit longitudinal or compression waves which have only one direction of vibration along each of the three orthogonal directions of propagation.) So for radiation $n = 6$ and $\gamma = 4/3$. Consequently, for a mixture of gas and radiation,

$$4/3 \leqslant \gamma \leqslant 5/3 \tag{56}$$

and the condition (52) for stability is satisfied.

It can be shown that γ decreases towards more massive stars. The total pressure is

$$P(\text{total}) = P(\text{gas}) + P(\text{radiation}) \tag{57}$$

From (11)

$$P(\text{radiation}) \propto T^4 \propto \frac{M^4}{R^4} \tag{58}$$

and from (8)

$$P(\text{total}) \propto \frac{M^2}{R^4} \tag{59}$$

consequently,

$$\frac{P(\text{total})}{P(\text{radiation})} = 1 + \frac{P(\text{gas})}{P(\text{radiation})} \propto M^{-2} \tag{60}$$

As M increases, P(radiation) increases relative to P(gas) and γ decreases towards $4/3$.

We have shown that the star is stable, and it is instructive to examine what happens to it if its equilibrium is slightly disturbed. Suppose that the radius changes slightly; as a consequence the total energy changes by a small amount ΔE where, from (51),

$$\Delta E = \frac{3\gamma - 4}{3(\gamma - 1)} \Delta\Omega \tag{61}$$

Now

$$\Omega = -\frac{qGM^2}{R} \tag{62}$$

where q is a numerical factor which depends on the relative distribution of mass within the star but is of order unity. Therefore,

$$\Delta\Omega = \frac{qGM^2}{R^2} \Delta R \tag{63}$$

and consequently it follows using (43) that ΔE has the same sign as ΔR if $\gamma > 4/3$. That is to say, if the radius decreases the star loses energy, which must be radiated away. But, from (50),

$$\Delta U = \frac{-\Delta\Omega}{3(\gamma - 1)} \tag{64}$$

and if $\gamma \geqslant 1$, ΔU has opposite sign to ΔR. Thus although the total energy decreases when R decreases, the internal energy increases. The source of this energy is gravitation; as the star contracts, and releases gravitational energy $\Delta\Omega$, a fraction

$$\frac{\Delta E}{\Delta \Omega} = \frac{(3\gamma-4)}{3(\gamma-1)} \tag{65}$$

is radiated away and a fraction

$$\frac{\Delta U}{\Delta \Omega} = \frac{-1}{3(\gamma-1)} \tag{66}$$

goes to increase the internal thermal energy, raising the temperature. The star becomes more luminous, and hotter. We have already seen (equation 11) that if we *assume* hydrostatic equilibrium and a perfect gas, the internal temperature varies as $T \propto 1/R$.

Note that, from (65) and (66), when $\gamma = 5/3$ (stars of low mass) the fraction of the gravitational energy radiated away is $\frac{1}{2}$ and consequently the fraction retained is $\frac{1}{2}$; for massive stars for which $\gamma \to 4/3$ the fraction radiated away $\to 0$ and almost all the gravitational energy released goes into increasing the internal energy. Conversely if a massive star expands a little the gravitational potential energy gains at the expense of the internal thermal energy.

The primary source of the luminosity of main sequence stars is, as we shall see, energy released by nuclear synthesis, and the effect of changes in the gravitational potential energy is relatively very small. This is not always the case in other stages of the evolution of a star which will be considered later in this book.

2.5. *The Equation of State.*

The equations of stellar structure formulated in section 2.3 require for their solution a specification of the equation of state of the material of which the star is composed, in the form of (40),

$$P = P(\rho, T, \text{composition})$$

In main sequence stars of moderate and high mass, the density is so low (figure 1) that at the temperatures involved the material behaves almost as a perfect gas, so that $P = P(\text{gas}) + P(\text{radiation})$, or

$$P = \frac{\rho \mathcal{R} T}{\mu} + \frac{aT^4}{3} \tag{67}$$

Since the gas in the stellar interior is completely ionised, the molecular weight μ is given by

$$\mu = (2X + \tfrac{3}{4}Y + \tfrac{1}{2}Z)^{-1} \tag{68}$$

where X, Y, Z are relative abundances by mass of hydrogen, helium and heavy elements.

On the other hand, in low mass main sequence stars the density becomes so high that the equation of state for a perfect gas is no longer even approximately valid. The properties possessed by a gas at very high densities can be examined as follows: by the Pauli Exclusion Principle, two similar particles in a volume $(\Delta q)^3$ at q cannot both have momenta in the range p to $p+\Delta p$, where

$$\Delta q \Delta p = h \qquad (69)$$

(Note that a pair of particles are not similar if they have opposite spins.)

When the gas density becomes very high, there are too many electrons per unit volume with similar momenta; curve 1 on figure 3.

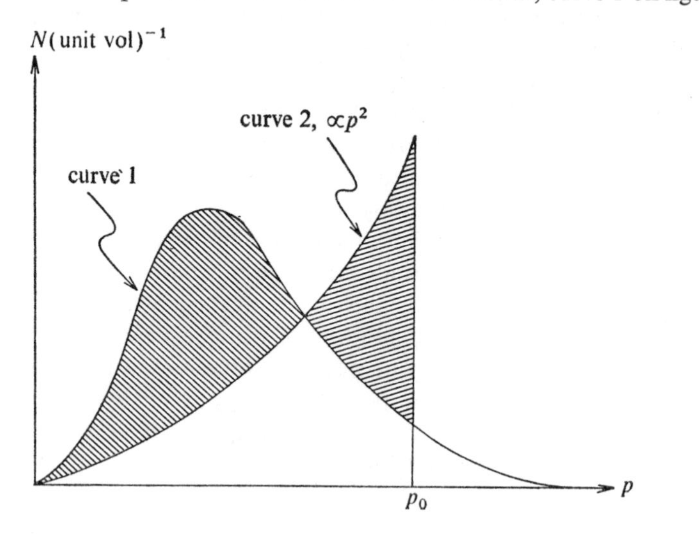

Figure 3. Showing the way in which the number N of particles having momentum p changes with momentum in a perfect gas (curve 1) and in degenerate matter (curve 2).

The momenta are redistributed by the operation of the exclusion mechanism. The distribution of momenta (and we remember that rate of transfer of momentum is synonymous with pressure) is then determined almost entirely by the density, and not by the temperature. The gas is degenerate.

The electron pressure becomes degenerate at much lower density than is required to make the ions degenerate. With equipartition of energy between ions and electrons $m_i v_i^2 = m_e v_e^2$, and it follows that the ratio of the momenta of the ions to the electrons is $(m_i/m_e)^{\frac{1}{2}}$. In the case of hydrogen, the main constituent of the star, $m_i/m_e = 1836$ and the ratio of the momenta is 43. Consequently, if the scale of curve 1 on figure 3 is regarded as applying to the electrons, the curve on the corresponding figure for hydrogen ions is stretched out along the abscissa by this factor and correspondingly reduced in height; the number density of ions at any momentum does not reach the limit permitted by the exclusion principle until this height is restored by a corresponding increase in the volume density of ions.

In the range of densities where electrons are degenerate and ions are not, the gas pressure is the sum of the ion pressure and the degenerate electron pressure; thus

$$P(\text{gas}) = P(\text{ions}) + P(\text{degenerate electrons}) \qquad (70)$$

where $P(\text{ions}) = n_i kT = \rho kT/\mu_i m_H$.

There are two alternative formulae for the pressure of a completely degenerate electron gas, which will be stated and then derived. If the kinetic energy of electrons $\ll m_e c^2$ (non-relativistic degeneracy, $\rho \leqslant 10^5$ g cm^{-3}), then

$$P(\text{d.e.}) = k_d \left(\frac{\rho}{\mu_e}\right)^{5/3} \qquad (71)$$

where $k_d = (3/8\pi)^{2/3} h^2 / 5 m_e m_H^{5/3}$ and μ_e is the number of proton masses per free electron.

Alternatively if the kinetic energy of the electrons $\not\ll m_e c^2$ (relativistic degeneracy, $\rho > 10^5$ g cm^{-3})

$$P(\text{d.e.}) = k_r \left(\frac{\rho}{\mu_e}\right)^{4/3} \qquad (72)$$

where $k_r = (1/8)(3/\pi)^{1/3}(hc/m_H^{4/3})$.

The derivation of (71) is as follows: on the basis that the system will take up the state of lowest energy we suppose that in each volume of the gas all the permitted momentum states are occupied up to some maximum value p_0. The volume of momentum space thus filled is

$$\int_0^{p_0} 4\pi p^2 \, dp$$

If we take unit volume of the gas, then from (69) with $\Delta q = 1$ the volume of a cell of momentum space which by the Exclusion Principle can contain only one particle of a given type is $(\Delta p)^3 = h^3$. The number of such cells available to be filled is therefore

$$\int_0^{p_0} \frac{4\pi p^2}{h^3} dp$$

Each cell can contain two electrons of opposite spins, and hence the total number of electrons in unit volume of the gas is

$$n_e = 2 \int_0^{p_0} \frac{4\pi p^2 \, dp}{h^3} = \frac{8\pi}{3} \frac{p_0^3}{h^3} \qquad (73)$$

Now pressure is the mean rate of transfer of momentum across unit area, and so

$$P_e = \frac{1}{3}\int_0^{p_0} (pv_e)\, dn_e \tag{74}$$

where v_e is the speed of the electron and the factor $\frac{1}{3}$ arises from averaging over all directions. Thus

$$P_e = \frac{1}{3}\int_0^{p_0} \left(\frac{p^2}{m_e}\right) 2\frac{4\pi p^2\, dp}{h^3}$$

that is,

$$P_e = \frac{8\pi}{15} \cdot \frac{p_0^5}{m_e h^3} \tag{75}$$

From (73) and (74) it follows that

$$\frac{P_e}{n_e^{5/3}} = \left(\frac{3}{8\pi}\right)^{2/3} \frac{h^2}{5m_e} \tag{76}$$

and since $\rho = n_e \mu_e m_H$ we have finally

$$P_e = \left(\frac{3}{8\pi}\right)^{2/3} \frac{h^2}{5m_e m_H^{5/3}} \left(\frac{\rho}{\mu_e}\right)^{5/3} \tag{77}$$

giving the equation of state $P_e(\rho)$ for a non-relativistically degenerate ionised gas.

The derivation of (72) for relativistic degeneracy proceeds in a similar manner, but the relativistic velocity must be used in (74); thus

$$\frac{v_e}{c} = \frac{p}{[(m_e c)^2 + p^2]^{\frac{1}{2}}} \tag{78}$$

which gives for the pressure

$$P_e = \frac{1}{3}\int_0^{p_0} (pv_e)\, dn_e = \frac{1}{3}\int_0^{p_0} \frac{cp^2}{[(m_e c)^2 + p^2]^{\frac{1}{2}}} \cdot 2 \cdot \frac{4\pi p^2}{h^3}\, dp$$

$$= \frac{8\pi}{3} \cdot \frac{c}{h^3} \cdot \int_0^{p_0} \frac{p^4\, dp}{[(m_e c)^2 + p^2]^{\frac{1}{2}}}$$

When the electron velocity is relativistic, $p \gg m_e c$, and the integral closely approaches p_0^4; hence

$$P_e = \frac{2\pi}{3} \cdot \frac{c}{h^3} \cdot p_0^4 \tag{79}$$

Equations (79) and (73) give

$$\frac{P_e}{n_e^{4/3}} = \frac{1}{8}\left(\frac{3}{\pi}\right)^{1/3} hc \tag{80}$$

and hence, using (77)

$$P_e = \frac{1}{8}\left(\frac{3}{\pi}\right)^{1/3}\left(\frac{hc}{m_H^{4/3}}\right)\left(\frac{\rho}{\mu_e}\right)^{4/3} \tag{81}$$

giving the equation of state $P = P(\rho)$ for a relativistically degenerate ionised gas.

The appropriate equation of state, (67), (77) or (81), is then used in the equation of hydrostatic equilibrium, (37), for solution of the equations of stellar structure. Such a solution however, also requires us to derive a detailed formula for the opacity of the material of the star.

2.6. *The Opacity.*

This will depend upon the frequency of the radiation, and on the density, composition and state of excitation and ionisation of the material through which the radiation is passing. To begin with we shall ignore the effect of frequency, and examine the mechanisms by which radiation is scattered and absorbed. There are four of these.

1. *Bound–bound Atomic Transitions.* The absorption or emission of radiation by an atom causes an electron in the atom to move from one bound energy state to another. Little energy is absorbed by such transitions because the differences in energy between the various bound states are small in comparison with photon energies in stellar interiors.
2. *Bound–free Atomic Transitions.* The removal of an electron from a bound energy state in an atom to a free state with kinetic energy of motion can absorb large photon energies, and this ionisation is the most important source of opacity at high densities – in dwarf (low mass) stars.
3. *Free–free Transitions.* The absorption of a photon by an electron moving in the field of an ion to which it is not bound.
4. *Scattering.* Collisions between photons and electrons scatter both components without loss of energy.

It should be remembered that there is no net loss of energy in any of these processes when a star is in an equilibrium state, as are main sequence stars. The material is neither heating nor cooling. What is happening can best be understood by recalling our earlier considerations in section 2.2(b), where it was shown that although the light travel time for a distance in vacuum equal to the radius of the Sun is only 2.3 seconds, the passage of radiation outwards in the Sun is so resisted by the material of which the Sun is composed that it takes about two million years for radiation to travel from the centre to the surface. When a photon is absorbed by an atom and subsequently re-emitted, its passage is delayed and its outward course deflected; when a photon is scattered its progress is diverted. The effect of opacity is to resist the flow of radiation and thereby to slow it down. It is analogous to electrical resistance, as was pointed out in section 2.2(b). The greater the opacity, the steeper the temperature gradient required to force a given flow of radiation through the material.

The detailed calculation of the coefficient of opacity for various gaseous compositions, as a function of temperature, density and frequency, is a lengthy process (the reader wishing to pursue this further may consult A.N. Cox, *Stellar Evolution*, edited by R.F. Stern and A.G.W. Cameron, p.123. New York: Plenum Press, 1966). We shall have to be content with a brief and elementary excursion into the problem but one which nevertheless illustrates the essential physical processes on which the more detailed calculations are based. We shall deal with bound–free absorption since this predominates at the higher densities present in the stars on the lower part of the main sequence.

In thermodynamic equilibrium, the rate of recombination of ions and electrons is equal to the rate of ionisation; that is to say, the rate of emission of radiation is equal to the rate of absorption. Consider what happens to individual photons and atoms. Photons have a net outward flow through the star; each time one is absorbed by an atom and ionises it, the ion recombines with a passing free electron and emits a photon in a random direction. When a parallel beam of radiation enters a volume of material, figure 4, part of it passes through untouched, and part is absorbed to be re-emitted in all directions. The amount δI which is re-emitted out of the forward direction will be proportional to the intensity of radiation incident, I, and to the amount of absorbing material in its path, $\rho\,\delta s$; that is,

$$\delta I = \kappa I \rho\, \delta s \qquad (82)$$

where κ is a constant of proportionality and is denoted the *mass absorption coefficient*. A similar argument applies to scattering.

To determine how the mass absorption coefficient itself depends on density, on temperature, and on the composition of the material, we proceed as follows. Since the rate of emission is equal to the rate of absorption it does not matter which we consider. If most of the atoms are neutral they are in an absorbing state and it is easier to consider absorption; conversely if most of the atoms are ionised and hence are in an emitting state. The latter is the case in stellar interiors. At $T = 10^7$ K, a temperature typical of the interior of a star, the kinetic energy of an electron, $\frac{3}{2}kT$, is 1.3 keV. Consequently electrons bound to atoms with potentials less than this will be knocked off by

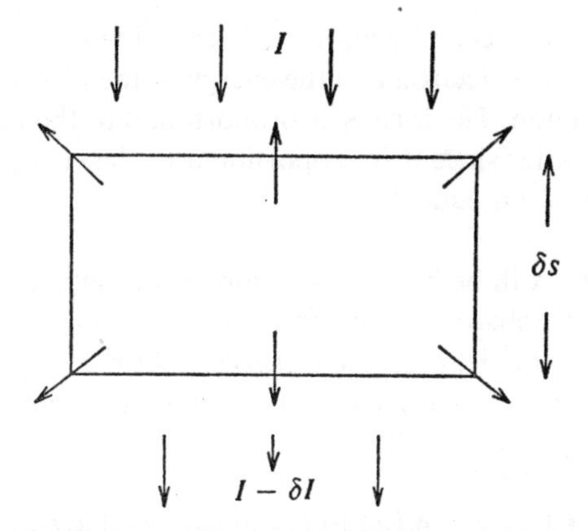

Figure 4. Radiation of intensity I entering a slab of gas of thickness δs is absorbed, re-emitted, and scattered in all directions, with the result that the ongoing intensity is reduced.

bombarding electrons; conversely, bombarding electrons are too energetic to be captured into levels with binding energies below 1.3 keV. Thus atoms will be stripped of electrons to their K shells.

For unit mass, the rate of emission of energy ($\propto$ rate of recombination) is proportional to: (a) the electron density, n_e; (b) the electron velocity, v_e; (c) the collison cross-section, σ; (d) the energy emitted per recombination, ε; and (e) the abundance of ions, n_i. Now

$$n_e = \sum_{\text{atoms}} (\text{atomic masses/unit volume}) \times (\text{electrons/atomic mass})$$

Let X denote the mass of hydrogen per unit mass of gas, Y and Z the corresponding quantities for helium and heavier elements. Then with m denoting unit atomic mass, N atomic number, and A atomic weight,

$$n_e = \left(\frac{\rho X}{m} \cdot 1\right) + \left(\frac{\rho Y}{4m} \cdot 2\right) + \left(\frac{\rho Z}{Am} \cdot N\right)$$

Noting that $N/A \approx \frac{1}{2}$ and in stars $Z \ll X$ so that $Y \approx 1 - X$, we obtain

$$n_e = \frac{\rho}{2m}(1+X) \tag{83}$$

The electron velocity is

$$v_e = \left(\frac{3kT}{m_e}\right)^{\frac{1}{2}} \tag{84}$$

and the collision cross-section is inversely proportional to the kinetic energy of the impinging electron,

$$\sigma \propto T^{-1} \tag{85}$$

The energy emitted on recombination, ε, is the difference between the kinetic energy of the electron and the energy of the level in which it is captured by the ion. The former is proportional to the temperature, and so is the latter since it is proportional to the energy of the ionising mechanism. Consequently,

$$\varepsilon \propto T \tag{86}$$

Most of the captures will be by ions with ionisation potentials large enough to capture bombarding electrons, which at the temperatures in stellar interiors, $\sim 10^7$ K, must have energies $\sim 10^3$ eV. That is to say, they will be ions of heavy elements, and so the relevant ion abundance has the proportionality

$$n_i \propto Z = (1 - X - Y) \tag{87}$$

Gathering these together, as in (a) to (e) above, we have:

$$\text{rate of absorption of energy} = \text{rate of emission of energy}$$
$$\propto \rho(1+X)(1-X-Y)T^{\frac{1}{2}} \tag{88}$$

The absorption coefficient κ can be described by $\kappa \propto$ rate of absorption of energy/flow of energy. Since flow of energy (by radiation transfer) $\propto T^4$ it follows that

$$\kappa = \kappa_0 \rho(1+X)(1-X-Y)T^{-3.5} \tag{89}$$

This is known as *Kramer's Opacity Law*, after its originator.

The determination of the constant of proportionality κ_0 involves, besides the distribution of electron velocities, the determination of collision cross-sections and energy emissions as functions of incident electron energy, kind of atom, and state of ionisation. It is of order 10^{25}.

It has been assumed above that the gas inside stars is highly ionised, and it was noted that atoms would be stripped of electrons to their K shells at temperatures of 10^7 K such as are found in stellar interiors. It is, however, possible to derive a formula giving the abundance of ions as a function of temperature and it will be useful to do so for future purposes as well as to put the above derivation of the Kramer's Opacity Law on to a firmer basis.

2.7. *The Level of Ionisation.*

Consider a Maxwellian distribution of velocities of electrons in a gas at 10^7 K, as illustrated in figure 5. The frequency distribution of kinetic energies peaks at about 10^3 eV. Ionisation potentials ψ of several atoms, e.g. H, He, O, are marked; electrons with energies less than ψ can be captured by the appropriate ions. Electrons with greater energies cannot be captured and are responsible for ionising the atoms to the energy levels indicated; for example, removing the K shell from oxygen. In practice, however, the kinetic energy of the impinging electron is shared with a bound electron, and the con-

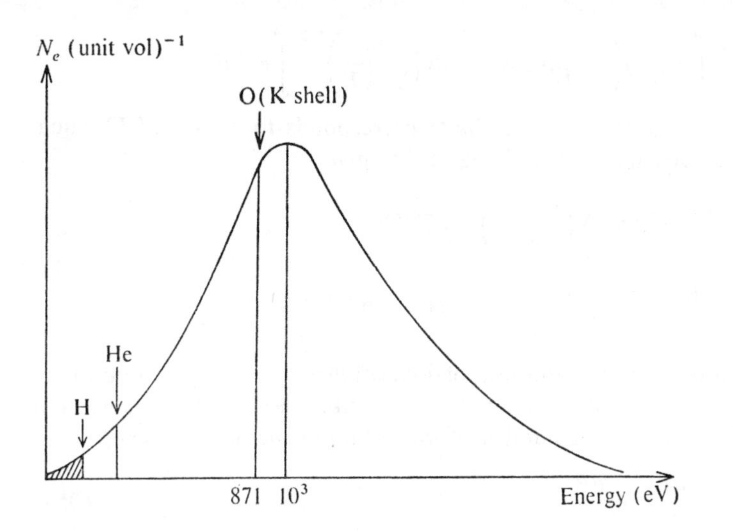

Figure 5. Kinetic energies of free electrons in a hot gas, compared to the binding energies of electrons in atoms.

dition for capture is KE < 2IP, that for ionisation, KE ⩾ 2IP. Consequently the rate of recombination collisions per ion for ions of a given kind is

$$\frac{dR}{dt} \sim \int_0^{\sqrt{(2\psi/m_e)}} \text{(number of electrons per unit velocity range)} \times \times \text{(velocity)} \times \text{(recombination cross-section)} \times \times \text{(velocity range)}$$

$$= \int_0^{\sqrt{(2\psi/m_e)}} N(v_e \pm \tfrac{1}{2}) v_e \left(\frac{a_0}{v_e^2}\right) dv_e \tag{90}$$

Similarly the rate of ionising collisions per atom is

$$\frac{dI}{dt} \sim \int_{\sqrt{(2\psi/m_e)}}^{\infty} N(v_e \pm \tfrac{1}{2}) v_e \left(\frac{a_0}{v_e^2}\right) dv_e \tag{91}$$

The Maxwellian distribution of velocities is given by

$$N(v)\,dv = 4\pi N \left(\frac{m}{2\pi kT}\right)^{3/2} e^{-mv^2/2\pi kT} v^2\,dv \tag{92}$$

Substituting $N(v_e)dv_e$ from (92) into (91) and writing

$$u = \left(\frac{m}{2\pi kT}\right) v^2; \qquad du = 2\left(\frac{m}{2\pi kT}\right) v\,dv \tag{93}$$

we have

$$\int N(v_e)\left(\frac{a_0}{v_e}\right)dv_e = 2\pi a_0 N\left(\frac{m}{2\pi kT}\right)^{1/2}\int e^{-u}\,du$$

and noting that $v_e = (2\psi/m_e)^{\frac{1}{2}}$ corresponds to $u = (\psi/\pi kT)$, equations (90) and (91) are integrated to give

$$\left.\begin{aligned}\frac{dR}{dt} &= 2\pi a_0 N\left(\frac{m}{2\pi kT}\right)^{1/2} e^{-\psi/\pi kT}\\ \frac{dI}{dt} &= 2\pi a_0 N\left(\frac{m}{2\pi kT}\right)^{1/2}(1-e^{-\psi/\pi kT})\end{aligned}\right\} \quad (94)$$

In equilibrium, the rate of ionisation is equal to the rate of recombination. So, if n_0 denotes the number of neutral atoms per unit volume and n_i the number of ionised atoms per unit volume,

$$n_0\frac{dI}{dt} = n_i\frac{dR}{dt} \quad (95)$$

That is, the ratio of ions to neutral atoms is

$$\frac{n_i}{n_0} = \frac{dI/dt}{dR/dt} = \frac{e^{-\psi/\pi kT}}{1-e^{-\psi/\pi kT}} \quad (96)$$

which gives the level of ionisation.

It is instructive to examine the level of ionisation for different ratios of ionisation potential ψ to the kinetic energy of the impinging electrons, using equation (96). Some values are shown in table 1; at 10^7 K, πkT is $\sim 10^3$ eV and it follows that $\psi/\pi kT$ is close to 0.01 for hydrogen, 0.05 for helium and 0.9 for oxygen; it can be seen from the table of values that whereas less than half the oxygen atoms are stripped to the K shell, hydrogen and helium are almost totally ionised.

$\psi/\pi kT$	n_i/n_0		$\psi/\pi kT$	n_i/n_0	
0.01	100	e.g. hydrogen	0.6	1.2	
0.05	20	e.g. helium	0.9	0.67	e.g. oxygen
0.1	9		1.5	0.28	
0.2	4.5		2.0	0.15	
0.4	2				

Table 1. Ionisation levels.

Since absorption of radiation by ionisation requires the presence of neutral or partially ionised atoms, it is apparent from table 1 that elements such as oxygen, although less abundant than hydrogen by a factor approaching a thousand, nevertheless play a major role in absorption, confirming the assumption made in writing (87).

For a more rigorous treatment we should take account of the distribution among the various possible degrees of ionisation of each kind of atom, which can be calculated by use of the 'Saha' equation. We could then determine the rate of recombination to each level, given the relevant cross sections, and hence the rate of emission of radiation. Equating this to the rate of absorption would give the absorption coefficient κ. Furthermore, we should also take account of the variation of opacity with the frequency of the radiation. To do this, we begin by assuming that, locally within the star, the energy flux and the temperature are related by the Planck function, then we can re-write (33) as a function of frequency:

$$F = \int_0^\infty F(\nu)\,d\nu = -\int_0^\infty \left[\frac{\frac{\partial}{\partial r}(\frac{4}{3}\pi B_\nu(T))}{\kappa_\nu \rho}\right] d\nu$$

$$= -\frac{4\pi}{3}\int_0^\infty \left[\frac{\frac{\partial B_\nu(T)}{\partial T}\cdot\frac{\partial T}{\partial \nu}}{\kappa_\nu \rho}\right] d\nu$$

$$= -\frac{4\pi}{3}\cdot\frac{1}{\rho}\cdot\frac{\partial T}{\partial r}\int_0^\infty \frac{\partial B_\nu(T)}{\partial T}\cdot\frac{1}{\kappa_\nu}\cdot d\nu$$

If we use a mean absorption coefficient κ given by:

$$\frac{1}{\kappa}\int_0^\infty \frac{\partial B_\nu(T)}{\partial T}\,d\nu = \int_0^\infty \frac{1}{\kappa_\nu}\frac{\partial B_\nu(T)}{\partial T}\,d\nu \tag{97}$$

we can use this mean κ directly in the relevant equations of stellar structure, (34) or (38). This is the so-called '*Rosseland*' *mean.* As (82) shows, it is a weighted mean where the weights are essentially reciprocals of the gradients of radiation pressure – the force driving the radiation outwards against the resistance of the material – at each frequency.

2.8. *Thomson Scattering by Free Electrons.*

In the more massive main sequence stars, the density is much lower (it will be recalled that $T_c \propto \mu M/R \sim$ constant along the main sequence, so that $M \propto R$ and $\rho \propto M/R^3 \propto M^{-2}$). At these lower densities the opacity due to bound-free transitions is correspondingly reduced, and that due to scattering of photons by electrons comes to dominate. As in (82), we have for the fractional loss of radiation $\delta I/I$ when passing through a thickness δs of the material,

$$\delta I/I = n_e \sigma\,\delta s = \kappa_e \rho\,\delta s$$

where σ is the scattering cross section of an electron,

$$\sigma = \frac{8\pi}{3}\left(\frac{e^2}{m_0^2}\right)^2 = 0.66 \times 10^{-24} \text{ cm}^2$$

Using (83) for n_e gives for κ_e the simple formula

$$\kappa_e = 0.2(1+X) \tag{98}$$

2.9. *Review.*

At this point it is well to pause to re-consider briefly the steps which we have taken and which have put us into the position of now being able to calculate the internal structure of main sequence stars.

The four basic equations required for this purpose were derived in section 2.3 and written in a convenient form in (36) to (39). They are equations which represent respectively the distribution of mass within the star, the balance of forces giving hydrostatic equilibrium, the outward flow of energy driven by the temperature gradient and resisted by the opacity, and finally the generation of energy within the star which continually replenishes that radiated away.

The internal pressure produced by the weight of the overlying layers increases towards the centre; the gas pressure must increase correspondingly to achieve the balance of forces for equilibrium, without which the star would not exist. This increase is obtained by inward increases in both temperature and density. Since the star is hot and is therefore radiating heat, its surface tends to cool and heat flows outward from the hotter interior. As the whole star thus tends to cool it contracts and releases gravitational energy: we showed by (66) that a fraction of this energy goes into increasing the internal temperature, and so we see that as a star radiates its heat away it becomes even hotter (in the next section we shall see that nuclear reactions provide an even more abundant source of energy from which the internal heat of main sequence stars is replenished). This increase in temperature and density with contraction increases the gas pressure and enables it to remain in balance with the pressure due to self-gravitation, that is, to support the weight of gas above any level. In order to do this the gas must have thermal properties which satisfy the criterion given by (52), that the ratio C_P/C_V should be greater than 4/3. If it were less than this value, reference again to (66) shows that the star would cool as it contracted, the gas pressure would not rise to balance the gravitation, and the star would collapse.

In order to solve (30) to (39), it is necessary to have formulae for the equation of state and the opacity of the stellar material, and these were derived in sections 2.5 to 2.8. Also, we need to know something about the energy source, but not very much; only, in fact, the distribution of the energy source within the star, to satisfy (39). This is rather a significant point. We do not have to know what *produces* the energy. For example, if we suppose that the energy is produced by nuclear reactions and (reasonably) that these depend steeply on the temperature of the material, we can assume that there is a point source of energy at the centre of the star without perhaps making too big an error. In this way, calculations of the internal structure of stars made considerable progress before the production of energy by nuclear synthesis was understood in any detail.

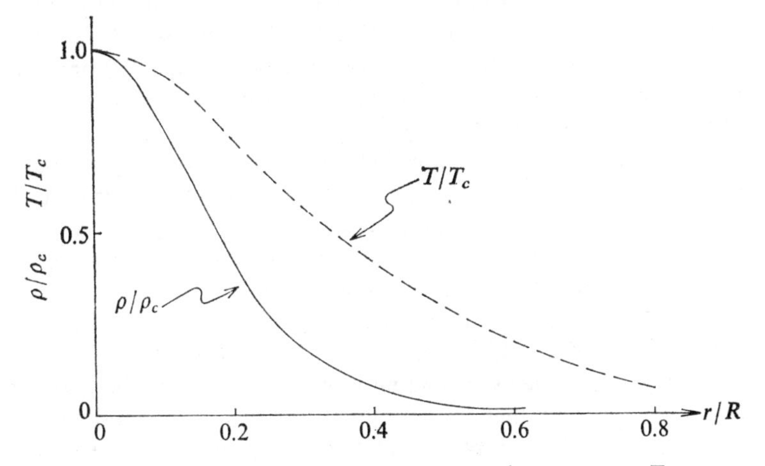

Figure 6. The radial distributions of density ρ and temperature T within a main sequence star of moderate mass.

The calculation of the internal structure of a star proceeds by numerical integration of (36) to (39), making use of the appropriate formulae for the equations of state, opacity and energy generations, and ensuring that the boundary conditions (43) are satisfied at the surface and at the centre. Typical results for the run of temperature and density with radius are shown in figure 6 for stars of intermediate mass, such as the Sun. We cannot carry through detailed calculations within the pages of a book such as this, but using (34) we can make a rough calculation of the average temperature inside the Sun. If we assume that the temperature gradient is linear, so that $dT/dr = T/R$, and use the opacity formula (89) with $\kappa_0 \approx 10^{25}$ as determined from laboratory data, then we have:

$$L = \frac{16\pi acRT^{7.5}}{3 \times 10^{25} \rho^2 (1+X)(1-X-Y)}$$

For the Sun, L is measured to be 4×10^{33} erg s^{-1}, $R = 7 \times 10^{10}$ cm,

$\bar{\rho} = 1.4$, and spectrochemical analysis gives $X = 0.7$ and $(1 - X - Y) = 0.02$; inserting these values gives $T^{7.5} \approx 10^{49}$. That is, the average temperature inside the Sun is $T \approx 3 \times 10^6$ K, from which we would expect that the central temperature $T_c \sim 10^7$ K.

2.10. *Energy Generation.*

Geological evidence shows that the age of the Sun must be about 5×10^9 yr or more and yet, as was shown in section 2.2, the total heat content is sufficient to last only 2×10^7 yr at the present rate of radiation, a rate that cannot have changed substantially over much of the geological time. There must therefore be some source of energy continually replenishing the internal heat of the Sun. During the lifetime of the Sun so far, this source must have generated an amount of energy $Lt/M \sim 3 \times 10^{17}$ erg g^{-1}.

Chemical energy, for example burning hydrogen in oxygen, would provide only some 3×10^{11} erg g^{-1}, and since only one per cent of the Sun's mass is oxygen, this is enough for only fifty years. A more likely possibility to be considered is the release of gravitational energy which would result from a slow contraction of the Sun. As each atom fell inwards, it would gain energy as it was accelerated by the gravitational field, until it collided with other atoms and so shared its energy among them in random motions, converting the gravitational energy into thermal energy of the gas. The gravitational energy released by contraction of the Sun from an infinite radius to its present size, which gives us an upper limit, is

$$E(\text{grav}) \sim GM/R = 2 \times 10^{15} \text{ erg g}^{-1}$$

which still falls short of requirements by a factor of more than a hundred.

However, if we turn to the energy released by fusion of atomic nuclei, a possible solution to the problem emerges.

(a) Fusion of hydrogen to form helium.

$4\,{}_1H^1 \rightarrow {}_2He^4$ releases $Q = 26.2 \text{ MeV}/4m_H = 6.3 \times 10^{18}$ erg g^{-1}

If the chemical composition of the radiating layers of the solar surface, as determined from spectrochemical analysis, is representative of the composition of the interior of the Sun, then about 80 per cent of the mass is hydrogen. Only a tenth of this would have had to be fused to helium to replenish the energy radiated during the Sun's lifetime.

(b) Fusion of helium to carbon.

$3\,{}_2He^4 \rightarrow {}_6C^{12}$ releases $Q = 7.4 \text{ MeV}/12m_H = 6 \times 10^{17}$ erg g^{-1}

The Sun contains about 17 per cent by mass of helium in its outer layers, and if the composition in the interior is similar, then only about 10^{17} erg g^{-1} is in practice available from this fusion reaction and this is too small by a factor of three.

(c) Fusion from carbon to iron. This releases about 10^{18} erg g^{-1}, but since the abundance of such heavier elements in the Sun is only about one per cent, the energy available is only $\sim 10^{16}$ erg g^{-1} of the solar material, very much less than the requirements.

It is concluded that of all the known sources of energy only that released by the fusion of hydrogen to helium is able to meet the Sun's needs. Since the Sun is a typical dwarf star, this may be supposed to be true of the other stars also. Consequently it is instructive to look in more detail at the process of nuclear fusion, because not only will it supply the required energy, it will also change the chemical composition of the star. Furthermore, for a precise solution of the equations of stellar structure, we need to know how the rate of fusion of nuclei depends on the composition, density and temperature of the material.

2.11. *Nuclear Synthesis.*

Several factors need to be considered: the collision cross section and the rate of collisions between nuclei, the probability of penetration of the potential barriers on collision, and the probability of transmutation, of fusion of the nuclei, when penetration occurs. These factors have to be determined for each of the different kinds of atomic nuclei and the different reactions in which they take part, and we also wish to calculate the energy released by the fusion processes.

a) *Probability of Penetration.* Consider two nuclei having electric charges $Z_1 e$ and $Z_2 e$ respectively, and separated by a distance r. The mutual potential energy ψ (figure 7) is given by

$$\psi = \frac{Z_1 Z_2 e^2}{r} \tag{99}$$

If the separation in the unexcited state is r_u and in the excited state it is r_e, and if the corresponding mutual potential energies are ψ_u and ψ_e, then the distribution of excited states is governed by

$$N(\psi_e) \propto e^{-\psi_u/\psi_e} \tag{100}$$

For interpenetration of approaching nuclei, the mutual kinetic energy of approach must exceed the mutual potential energy due to repulsion by the like electric charges in the excited state; that is,

$$E \geqslant \psi_e \tag{101}$$

Hence the probability of penetration of the potential barrier is

$$P(\mathrm{P}) = e^{-\psi_u/E} \tag{102}$$

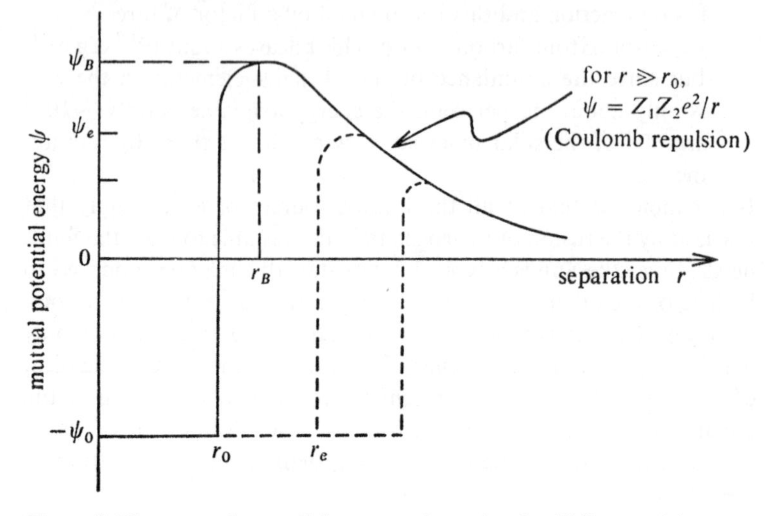

Figure 7. The mutual potential energy of a pair of colliding nuclei.

The mutual kinetic energy is

$$E = \frac{p^2}{2m} \tag{103}$$

where p is the momentum and m the reduced mass

$$m = \frac{m_1 m_2}{(m_1+m_2)} \tag{104}$$

If and when the colliding particles have interpenetrated they are no longer distinguishable as separate particles: the uncertainty in the position of either particle is r_e, the separation of the approaching particles at their maximum repulsion in the excited state; and their mutual momentum will have some indeterminate value in the range zero to p. It follows from Heisenberg's Uncertainty Principle that

$$pr_e \sim \frac{h}{2\pi} = \hbar \tag{105}$$

and consequently using (103) and (105)

$$E = E^{\frac{1}{2}} \, . \, E^{\frac{1}{2}} = \frac{hE^{\frac{1}{2}}}{2\pi r_e (2m)^{\frac{1}{2}}} \tag{106}$$

The probability of penetration becomes, using (99), (102) and (106),

$$P(\mathrm{P}) = e^{-(Z_1Z_2e^2/r_u)/(hE^{\frac{1}{2}}/2\pi r_e(2m)^{\frac{1}{2}})} \tag{107}$$

and noting that $r_e/r_u = \psi_u/\psi_e \geqslant \psi_u/E = \log_e(1/P)$ we can write without much error

$$P(\mathrm{P}) = e^{-2\pi(2m)^{\frac{1}{2}} \log_e(1/P)(Z_1 Z_2 e^2/hE^{\frac{1}{2}})} \tag{108}$$

A more rigorous treatment results in the factor $\log_e(1/P)$ being replaced by π, and the addition of a term depending on the separation r; thus (Gamow and Teller 1938, *Physical Review* **53**, 608)

$$P(\mathrm{P}) = e^{-(\alpha-\beta)} \tag{109}$$

where $\alpha = 2\pi^2(2m)^{\frac{1}{2}} Z_1 Z_2 e^2/hE^{\frac{1}{2}}$ and $\beta = 8\pi(2m)^{\frac{1}{2}}(Z_1 Z_2 e^2 r)^{\frac{1}{2}}/h$. Note that in both (108) and (109), since the exponent contains a factor $m^{\frac{1}{2}}Z^2$ the probability of penetration *decreases rapidly with increasing atomic number.*

b) *Rate of Collisions.* Consider nuclei of two types with number densities N_1 and N_2 per unit volume; let v and E denote the velocity and energy of relative motion, σ the cross section for mutual collision and m the reduced mass. The rate of collisions per unit volume between the two types of nuclei with total velocities of relative motion in the range v to $v+dv$ is

$$\frac{dN}{dt} = \sigma N_1 N_2(v) v\, dv \tag{110}$$

or, since $2E = mv^2$,

$$\frac{dN}{dt} = \frac{\sigma}{m} N_1 N_2(E)\, dE \tag{111}$$

For a Maxwellian distribution of velocities (see text books on kinetic theory),

$$N_2(v)\, dv = 4\pi N_2 \left(\frac{m}{2\pi kT}\right)^{3/2} e^{-mv^2/2\pi kT} v^2\, dv \tag{112}$$

or, in energy terms,

$$N_2(E)\, dE = 8\pi N_2 \frac{m^{\frac{1}{2}}}{(2\pi kT)^{3/2}} e^{-E/\pi kT} E\, dE \tag{113}$$

It follows from (111) and (113) that

$$\frac{dN}{dt} = \frac{8\pi N_1 N_2}{(2\pi kT)^{3/2}} \cdot \frac{\sigma}{m^{\frac{1}{2}}} \cdot e^{-E/\pi kT} E\, dE \tag{114}$$

c) *Collision Cross Section.* As in 2.11(a) above, when two particles with mutual momentum p have approached sufficiently close to be indistinguishable, they have collided and again we have $pr_e = \hbar$. That is to say, collision requires an approach within a distance r_e so that the cross section for collision is

$$\sigma = \pi r_e^2 = \frac{\pi \hbar^2}{p^2} = \frac{\pi \hbar^2}{2mE} \tag{115}$$

d) *Probability of Transmutation per Penetration.* We shall use the concept of the compound nucleus introduced by Niels Bohr. It is supposed that when two nuclei interpenetrate their energy of collision is shared among their constituent nuclei, forming a compound nucleus in an excited state. If this holds together long enough it will emit the excitation energy as electromagnetic energy and form a stable unexcited nucleus; otherwise it will emit one or more particles.

The excited state will have a number of energy levels; let the sum of their widths be ΔE. The time τ which the compound nucleus spends in the excited state will, by the Uncertainty Principle, be

$$\Delta E\tau = \hbar \tag{116}$$

The corresponding momentum width is Δp, and denoting the radius of the compound nucleus by r_e (giving the uncertainty in position of the components) it follows that

$$\Delta p r_e = \hbar \tag{117}$$

The longer the compound nucleus survives, the greater the chance of transmutation; that is, the probability of transmutation

$$P(\mathrm{T}) \propto \tau \tag{118}$$

Denote by t the length of time for which the compound nucleus must survive in order to emit electromagnetic radiation and become a new bound nucleus; then

$$P(\mathrm{T}) = \frac{\tau}{t}, \qquad \tau \ngtr t. \tag{119}$$

Since $\Delta E = (\Delta p)^2/m$, then using (116), (117) and (119) and writing $\Gamma = 1/t$ we have

$$P(\mathrm{T}) = \Gamma \frac{m r_e^2}{\hbar} \tag{120}$$

e) *Rate of Transmutations.* We can now calculate the rate of transmutations per unit mass; this is

$$\frac{dN(\text{trans})}{dt} = \frac{dN}{dt} P(\mathrm{P}) P(\mathrm{T}) \frac{1}{\rho} \tag{121}$$

and using (109), (114), (115) and (120) this becomes

$$\frac{dN(\text{trans})}{dt} = 5.3 \times 10^{25} \rho \left(\frac{X_1}{A_1}\right)\left(\frac{X_2}{A_2}\right) \Gamma \Sigma \phi^2 e^{-\phi} \tag{122}$$

where:

X_1, X_2 are the relative abundances by mass of the two types of nuclei of atomic masses A_1 and A_2 respectively;

$\Sigma = (r_0^{\frac{3}{2}}/(AZ_1Z_2)^3)\exp(2r_0^{\frac{1}{2}})$ is an effective cross-section;

$r_0 = (8r/a_0)$; $a_0 = \hbar^2/mZ_1Z_2e^2$ is the Bohr radius;
$A = A_1A_2/(A_1+A_2)$ is the reduced atomic mass; and
$\phi = \phi_0(AZ_1^2Z_2^2/T)^{1/3}$ measures the relative energy of collision in terms of the Coulomb barrier. $\phi_0 = 4248$ if $\Gamma\hbar$ is given in volts (note that $\Gamma\hbar$ has units of energy and is called the *width* of the reaction). The cross-section is sometimes written as S_0 where

$$S_0 = \frac{\pi}{2} \cdot \frac{\Gamma\hbar(a_0/8)^2}{(AZ_1Z_2)^3}\Sigma$$

f) *Rate of Energy Generation.* Let Q denote the energy liberated per transmutation. Then the rate of generation of energy per unit mass is

$$\varepsilon = Q\frac{dN(\text{trans})}{dt} \tag{123}$$

With these equations we are now able to examine the process of synthesis of hydrogen to helium, which as we have already noted is the only reaction able to supply sufficient energy to replenish that radiated by the stars; and the dependence of the rate of energy generation on temperature and density can be determined to be used in (39) for a solution to the equations of stellar structure.

2.12. *The Proton–Proton (pp) Chain Reaction.*
This begins with the transmutation of hydrogen to deuterium:

a) ${}_1H^1 + {}_1H^1 \rightarrow {}_1H^2 + \beta^+ + \nu$

The neutrino quickly escapes from the star and the positron combines with an electron to give gamma radiation, $\beta^+ + \beta^- \rightarrow \gamma$, which feeds the radiation field in the star. Then hydrogen and deuterium react to form an isotope of helium:

b) ${}_1H^2 + {}_1H^1 \rightarrow {}_2He^3 + \gamma$

Collisions between He^3 nuclei then give He^4:

c) ${}_2He^3 + {}_2He^3 \rightarrow {}_2He^4 + 2{}_1H^1$

The net effect of the chain is to fuse four hydrogen nuclei into one helium nucleus. The values determined for the cross section, energy per transmutation, and mean lifetimes are shown in table 2.

Reaction	$S_0(10^{-24}$ keV cm$^2)$	Q (MeV)	$1/\Gamma$ (s)
(a)	2.9×10^{-22}	1.443	3×10^{17}
(b)	1.0×10^{-4}	5.494	6
(c)	2×10^3	12.847	3×10^{13}

Table 2. The reactions of the *pp* cycle.

The energy released in the first reaction includes 0.257 MeV for the energy carried by the neutrino. Excluding this, since it escapes from the star, the total energy released in $4{}_1H^1 \rightarrow {}_2He^4$ is

$$Q = 2(Q_a + Q_b) + Q_c = 26.207 \text{ MeV} \tag{124}$$

The overall rate is governed by the slowest reaction in the chain which, on account of the small cross-section S_0, is reaction (a); hence the rate of energy generation ε is calculated from (123) using Q from (124), with S_0 and Γ in (122) taking the values for reaction (a), and r_0 referring to the radius of the hydrogen atom.

The only parameters in (122) and (123) not determined are the density, temperature and hydrogen abundance, and so the rate of energy generation can be written

$$\varepsilon(pp) = \varepsilon(pp)_0 \rho X^2 f(T) \tag{125}$$

noting that $X_1 = X_2 = X$, the hydrogen abundance. It is convenient in calculations of stellar structure to represent $f(T)$ by a power formula approximation,

$$f(T) = T^n \quad \text{where } n = \frac{d \log_e \varepsilon}{d \log_e T} \tag{126}$$

Calculation of ε from (122) and (123) in the form of (125) then gives rates which can be closely represented by (126), with n being found to have the values shown in table 3. The values for ε_0 are for T in units of 10^7 K in equation (125) and ε has been calculated for $\rho = 100$ g cm^{-3} (the value at the centre of the Sun) and a hydrogen abundance $X = 0.67$. Note that the mean energy production by the Sun is 2 erg g^{-1} s^{-1}.

T 10^7 K	$\varepsilon(pp)_0$	n	$\varepsilon(pp)$ erg g^{-1} s^{-1}
0.5	0.09	6.0	0.1
1.0	0.06	4.5	3
1.5	0.07	4.0	17
2.0	0.09	3.5	
2.5	0.11	3.25	

Table 3. Values of n in equation (126).

Solution of the equations of stellar structure show that the central temperatures of dwarf main sequence stars are about 10^7 K, at which temperature the *pp* reaction considered above is a much more efficient process than the hydrogen fusion cycle involving carbon and nitrogen to be studied in the next section.

Consider a dwarf star with the Kramer's opacity law; from (14), remembering that $L = 4\pi R^2 H$, $T_c \propto M/R$, $\rho \propto M/R^3$ and $T \propto T_c$ for a given structure, we have for the energy outflow

$$L = C_1 M^5 T_c^{0.5} \tag{127}$$

But from (39), (125) and (126), the energy production is

$$L = \bar{\varepsilon}M \propto \rho M T^4 = \frac{C_2 T_c^7}{M} \tag{128}$$

For equilibrium, the right-hand sides of (127) and (128) must be equal, from which it follows that

$$T_c = (C_1/C_2)^{1/6.5} M^{0.9} \tag{129}$$

Note that (127) can then be written

$$L \propto M^{5.45} \tag{130}$$

giving a mass-luminosity relation for dwarf main sequence stars.

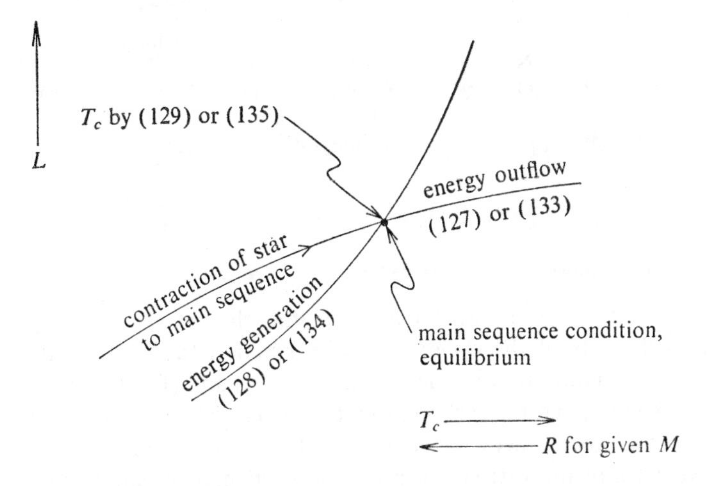

Figure 8. Showing how energy generation and energy outflow depend on the central temperature in a star. The point of intersection represents the equilibrium condition which holds on the main sequence.

We can best see how the balance between luminosity and energy generation is achieved by graphical means, figure 8. As the star contracts towards the main sequence state, its central temperature and energy outflow increase slowly. The energy outflow exceeds the energy produced by the nuclear reactions and the deficit is made up by the release of gravitational energy – which means that the star has to continue to contract: we have already noted in (66) that only part of the energy so released is radiated, the remainder heating up the stellar interior. This state of affairs continues until the production of nuclear energy, which has been rising very steeply with temperature, equals the outflow. The star remains in that equilibrium condition until the supply of nuclear energy is exhausted.

2.13. *The Carbon–Nitrogen* (CN) *Cycle.*

This cycle utilises carbon, nitrogen and oxygen in a series of reactions which convert hydrogen to helium, and at the slightly higher temperatures in giant as against dwarf stars, they do so with greater efficiency than the *pp* cycle. The series of reactions are given in table 4, together with data on cross sections, energy release and mean lifetimes.

	Reaction	S_0 (10^{-24} keV cm^2)	Q (MeV)	$1/\Gamma$ (s)
(a)	${}_6C^{12}+{}_1H^1={}_7N^{13}+\gamma$	1.2	1.945	3×10^{14}
(b)	${}_7N^{13}\rightarrow {}_6C^{13}+\beta^++\nu$	disintegration	1.502 (+0.72 neutrino)	420
(c)	${}_6C^{13}+{}_1H^1={}_7N^{14}+\gamma$	6–12	7.542	3×10^{13}
(d)	${}_7N^{14}+{}_1H^1={}_8O^{15}+\gamma$	3 (non-resonant) 30 (resonant)	7.347	3×10^{15}
(e)	${}_8O^{15}\rightarrow{}_7N^{15}+\beta^++\nu$	disintegration	1.729 (+0.976 neutrino)	82
(f)	${}_7N^{15}+{}_1H^1={}_6C^{12}+{}_2He^4$	10^5	4.961	3×10^{12}

Table 4. The reactions of the CN cycle.

If reaction (d) is non-resonant it is the slowest reaction and controls the rate of the cycle; otherwise reaction (a) controls the rate. Hydrogen atoms are added at (a), (c), (d) and (f) to give one atom of helium, and the carbon atom used in (a) is recovered in (f). The relative abundances of N, O and C are determined by the relative rates of the various reactions; in particular, because (d) is slow a large abundance of ${}_7N^{14}$ builds up.

If (d) controls the rate, then (122), (123) and (125) give

$$\varepsilon(CN)=\varepsilon(CN)_0\rho XZ(N^{14})T^n \tag{131}$$

where n is defined by (126). On the other hand if the rate is controlled by (a), then

$$\varepsilon(CN)=\varepsilon(CN)_0\rho XZ(C^{12})T^n \tag{132}$$

Calculation as before gives the values for $\varepsilon(CN)_0$ and n against temperature shown in table 5.

T 10^7 K	$\varepsilon(CN)_0$	n
0.5	3×10^{-3}	29
1.0	3.5×10^{-4}	23
1.5	1.6×10^{-4}	20
2.0	1.8×10^{-3}	18
2.5	4.8×10^{-3}	16.6

Table 5.

Consider a massive star, a main sequence giant, for which from (20) the energy outflow is

$$L=C_3M^3 \tag{133}$$

As before from (39) with (131) or (132), we have for the energy generation

$$L=\bar{\varepsilon}(\mathrm{CN})M \propto \rho M T^{23} = \frac{C_4 T_c^{26}}{M} \tag{134}$$

For equilibrium the right hand sides of (133) and (134) must agree (see figure 8): hence

$$T_c = \left(\frac{C_3}{C_4}\right)^{1/26} M^{0.15} \tag{135}$$

The relative rates of energy production by the *pp* and CN cycles can be determined for various temperatures from (125), (126) and (131) or (132);

$$\frac{\varepsilon(\mathrm{CN})}{\varepsilon(pp)} = \frac{\varepsilon(\mathrm{CN})_0}{\varepsilon(pp)_0} \cdot \frac{Z(\mathrm{CN})}{X} \cdot T^{n(\mathrm{CN})-n(pp)} \tag{136}$$

Assuming $Z(\mathrm{CN})/X = 0.005$, table 6 shows the results obtained by inserting the tabulated values in (136).

$T(10^7\ \mathrm{K})$	$\varepsilon(\mathrm{CN})/\varepsilon(pp)$	$T(10^7\ \mathrm{K})$	$\varepsilon(\mathrm{CN})/\varepsilon(pp)$
0.5	3×10^{-11}	1.8	1.0
1.0	4.5×10^{-5}	2.0	3.3
1.5	1.6×10^{-2}	2.5	84

Table 6.

The fact that the central temperatures of stars increase slowly with mass shows that the *pp* cycle dominates in dwarf stars, and the CN cycle in massive ones. The relationships between mass, luminosity and central temperature are illustrated in figure 9. At small masses the *pp* cycle predominates. Going up the main sequence to larger masses, the luminosity L increases as $M^{5.45}$, and the central temperature T_c increases as $M^{0.9}$ so that the nuclear energy generation keeps pace with the luminosity (note that the density decreases up the main sequence, but the temperature change has a much larger effect on the rate of nuclear synthesis). At these central temperatures, the energy produced by the CN cycle is much less than that produced by the *pp* cycle, but it is increasing much more quickly with increasing temperature and when $T_c = 1.8\times10^7$ K the rates of energy production by the two cycles are equal. This is in stars of about twice the mass of the Sun. Above $2M_\odot$, the CN cycle rapidly takes over essentially all the energy production, and the central temperature rises less quickly with mass, $T_c \propto M^{0.15}$, since $\varepsilon(\mathrm{CN})$ is much more temperature sensitive than $\varepsilon(pp)$. Note that the change from

Kramer's opacity to electron scattering occurs at a higher mass than $2M_\odot$, so that the luminosity continues to increase with mass as $M^{5.45}$ after the CN cycle has taken over, then gradually changes to $L \propto M^3$ for higher masses.

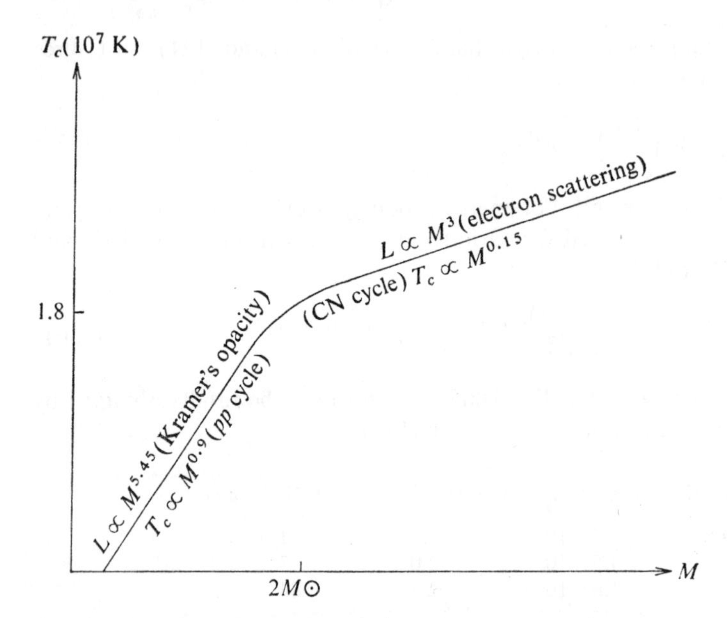

Figure 9. The dependence of central temperature on stellar mass, showing also how the dependence of luminosity on mass changes on account of changes in the opacity law.

In stars with masses greater than $1.7M_\odot$ the CN cycle contributes significantly to the energy generation, and the steep temperature dependence of ε(CN) results in a very high proportion of the energy production occurring in the centre of the star. Radiation alone is unable to transport this energy outwards rapidly enough, and the temperature gradient increases due to the build-up of the heat content close to the centre until dT/dP exceeds the limit for radiative equilibrium which was derived in (27). Convection ensues, producing a convective core in the star which extends outwards from the centre until it reaches a large enough surface area to enable radiative transport to take over without the temperature gradient becoming excessively steep. The size of the convective core increases with increasing mass, because the energy outflow increases so much more quickly than the mass; in stars with masses of $30M_\odot$ it encompasses half the mass of the star.

A convection zone having a different origin occurs in the outer layers of dwarf stars. The equation for energy flow (equation 14) can be re-written

$$\frac{dT}{dP} \propto \frac{\kappa H}{\left(\dfrac{T^3}{\rho}\dfrac{dP}{dr}\right)} \qquad (137)$$

dT/dP will increase if either the opacity coefficient κ or the energy outflow H is increased, and if it exceeds the limit set by (27) convection ensues. We have seen above a case where convection results from a very high value for H; in the outer layers of dwarf stars hydrogen is only partially ionised and a very high level of opacity results, causing dT/dP to exceed the critical value for convection. In attempting to force the radiation through the highly opaque gas, the temperature gradient has become too steep for radiative equilibrium.

2.14. *The Rate of Gravitational Contraction to the Main Sequence.* It was noted in discussing figure 8 that before a balance is reached between the energy outflow and nuclear energy production in a star – the long-lasting equilibrium situation which produces the main sequence – the star progressively contracts, releasing gravitational energy. The gravitational potential energy is $\Omega = -qGM^2/R$, where q depends on the interior distribution of mass but is of order unity. Now $L = \Delta E/\Delta t$ and by (65) this becomes

$$L = \frac{3\gamma-4}{3(\gamma-1)}\frac{\Delta\Omega}{\Delta t} = \frac{3\gamma-4}{3(\gamma-1)}q\frac{GM^2}{R^2}\frac{dR}{dt} \qquad (138)$$

The equation of stellar structure (39) becomes

$$\varepsilon(\text{grav}) = \frac{3\gamma-4}{3(\gamma-1)}q\frac{GM}{R^2}\frac{dR}{dt} \qquad (139)$$

Solution of the equations of stellar structure then gives a value for dR/dt. That is to say, the star must contract at such a rate that the release of gravitational energy given by (138) equals the outflow of radiation given by (38). For example, if energy is produced too quickly, the star will heat up, the internal pressure will rise and contraction will be slowed down until the energy output is not excessive.

Note that in all the cases we have considered, the structure of the star is not greatly affected by the source of energy generation; it is determined primarily by the condition of hydrostatic equilibrium, and by the properties of radiative transfer of energy.

Hydrostatic equilibrium requires that gas pressure balances gravitational self-attraction; since the weight of the overlying layers increases inwards, density and temperature must both increase inwards. By the first law of thermodynamics this causes an outflow of radiation; thus it is the condition of hydrostatic equilibrium that causes the star to be luminous. The opacity, which restricts the outflow of radiation, prevents the star from cooling catastrophically and thus prolongs the equilibrium condition; it determines *how* luminous the star will be.

The source of energy generation, by replenishing the energy radiated away, makes the star stable. Because particular sources of nuclear energy operate at the required rate at particular temperatures, they stabilise a star at particular appropriate radii ($R \propto M/T_c$) and hence at particular luminosities ($L = L(M,R)$) and effective temperatures ($T_e = T_e(L,R)$); these particular stable states are represented by the main sequence, red giant sequence, etc.; any concentration of points on the L, T_e diagram represents a relatively long lifetime in the corresponding state.

Thus the law of gravity and the opacity of the gas are the dominating factors governing the structure, the distribution *within* the star of the mass and the temperature. The star adjusts its *overall* radius so that the central temperature takes up just the value required for nuclear synthesis to produce energy as fast as the star is radiating it away; or, if there is no source of nuclear energy in the conditions of temperature and density existing in the star, it contracts at a rate which it adjusts to release gravitational energy just quickly enough to balance the radiation.

2.15. *Main Sequence Lifetimes.*

Not all the mass of a star is available as nuclear fuel on the main sequence. Most of it is too far from the centre to reach the high temperatures required for nuclear reactions. Even where convection occurs in the more massive stars, in which the CN cycle operates at the centre, it is confined to the core of the star and consequently most of the mass is not mixed by convection into the central part.

The energy supply from the fusion of hydrogen to helium is, as we have seen, 6×10^{18} erg g^{-1}. If the mass of fuel available is denoted M_c, the main sequence lifetime is

$$t(\mathrm{MS}) = \text{energy supply}/\text{luminosity}$$

$$= 6 \times 10^{18} \frac{M_c}{L} = 6 \times 10^{18} \frac{M_c}{M} \cdot \frac{M}{L} \tag{140}$$

For example, in the case of a star of solar mass, $M_c/M \approx 0.1$, and $M/L = 0.5$; hence

$$t(\odot) = 3 \times 10^{17} \text{ s} = 10^{10} \text{ yr}$$

This lifetime is about twice the geological age of the earth and consequently the Sun is a middle-aged star.

It is a very different story in the case of massive stars. A star of about $50M_\odot$ has $M_c/M \approx 0.5$ and $M/L = 5 \times 10^{-4}$ (figure 1). Hence

$$t(\mathrm{MS}, 50M_\odot) = 1.5 \times 10^{15}\ \mathrm{s} = 5 \times 10^{7}\ \mathrm{yr}$$

very much less than the age of the earth. Massive stars formed more than 10^8 yr ago will have completed their main sequence lifetimes; what form will they have taken?

3. Red Giants. 3.1. 3.2. *The Evolution of a Star of Mass* $5M_{\odot}$.

3.1.

When all the hydrogen in the core of a main sequence star has been synthesised to helium, the supply of energy from the proton–proton or carbon–nitrogen cycles is no longer available. It has been noted before, in considering figure 8, that where the energy available from nuclear synthesis is inadequate to replenish the outflow of energy from the star, the deficiency is made up from gravitational energy; the star steadily contracts. The contraction begins as the star attempts to increase the energy output from a diminishing abundance of hydrogen; since

$$L = \varepsilon M = \varepsilon_0 M \rho X^2 T_c^n$$

T_c rises as X falls, and of course $R \propto M/T_c$. The contraction releases gravitational energy, which takes over an increasing share of the energy supply as X diminishes towards zero; consequently T_c increases only slowly.

However, the composition is now non-uniform, figure 10. The boundary conditions which we applied to the solution of (36) to (39) for main sequence stars referred only to homogeneous stars

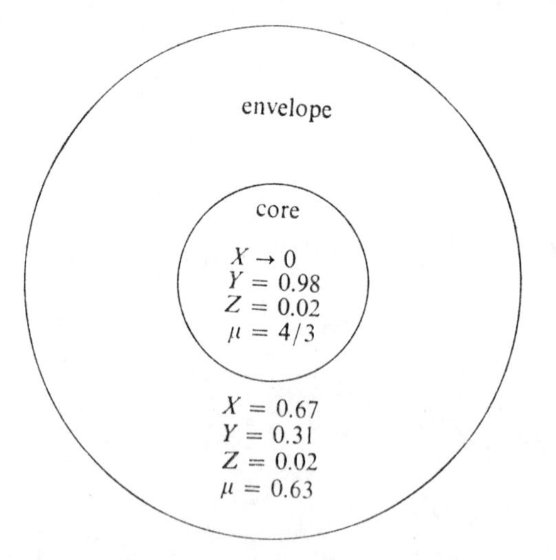

Figure 10. The interior composition of a star when it has synthesised all the hydrogen in the core to helium. This marks the end of the main sequence phase of its life.

and are not adequate for this new circumstance. More conditions must be imposed. At the discontinuity in molecular weight, the boundary between core and envelope:

1. the pressure cannot be discontinuous, because a pressure discontinuity would represent a shock wave which would travel through the gas to remove the discontinuity;
2. the temperature cannot be discontinuous since an infinite temperature gradient would imply an infinite energy flow and hence an infinite luminosity;
3. the outflow of radiation must be continuous – there must be no sink or source of L at the boundary (unless there is a shell source of nuclear energy);
4. the mass must be continuous – there must be no shell of infinite density.

In other words, if suffix c represents the core and e the envelope, at the boundary $r = r_c$

$$P_e = P_c, \ T_e = T_c, \ L_e = L_c \text{ (for no shell source)}, \ M_c = M_e \quad (141)$$

(T_c is not to be confused with the central temperature; M is the mass interior to the radius.)

We now enquire as to what happens to the interior structure of the star; in particular, what happens to the density *gradient* at the core boundary? Furthermore, since the nuclear fuel in the core is exhausted, what would be the effect of energy generation in a shell at the core boundary, the hottest region with unsynthesised hydrogen fuel left?

Assuming that the gas is not degenerate and that radiation does not dominate the pressure, the equation of state is

$$P = \frac{\rho \mathcal{R} T}{\mu} \quad (142)$$

It follows that

$$\frac{\rho_c}{\rho_e} = \frac{\mu_c}{\mu_e} \quad (143)$$

In the example given above, $\mu_c / \mu_e = 2$. Now from (34),

$$L \propto \frac{r^2 T^3 (dT/dr)}{\kappa \rho}$$

Suppose that there is a shell source of energy; write

$$L = L_e = (1+u) L_c$$

where L is the total energy outflow, L_c is the energy flowing out of the core, and uL_c is the energy generated in a shell source at the core boundary; note that

$$L(\text{shell}) = uL_c = \frac{uL}{(1+u)} \tag{144}$$

It follows that

$$\left(\frac{dT/dr}{\kappa\rho}\right)_e = (1+u)\left(\frac{dT/dr}{\kappa\rho}\right)_c$$

Since $T_c = T_e$, any dependence on temperature of the opacity coefficient κ is irrelevant at the boundary. The dependence of κ on density ρ and on hydrogen abundance X is represented by

$\kappa \propto (1+X)\rho^n$

where $n = 0$ for electron scattering and $n = 1$ for Kramer's opacity. Then

$$\frac{(dT/dr)_e}{(dT/dr)_c} = (1+u)\frac{(1+X_e)\,\rho_e^{n+1}}{(1+X_c)\,\rho_c^{n+1}}$$

$$= (1+u)\frac{1+X_e}{1+X_c}\left(\frac{\mu_e}{\mu_c}\right)^{n+1} \tag{145}$$

Also, since $M_c = M_e$ it follows from the equation of hydrostatic equilibrium (31) that $dP/dr \propto \rho$ at $r = r_c$, and hence, using (143):

$$\frac{(dP/dr)_e}{(dP/dr)_c} = \frac{\mu_e}{\mu_c} \tag{146}$$

Differentiating (142),

$$\frac{dP}{dr} = \frac{\mathcal{R}}{\mu}\left(T\frac{d\rho}{dr} + \rho\frac{dT}{dr}\right)$$

or

$$\frac{d\rho}{dr} = \frac{1}{T}\left(\frac{\mu}{\mathcal{R}}\frac{dP}{dr} - \rho\frac{dT}{dr}\right)$$

Hence,

$$\frac{\left(\frac{d\rho}{dr}\right)_e}{\left(\frac{d\rho}{dr}\right)_c} = \frac{\frac{\mu_e}{\mathcal{R}}\left(\frac{dP}{dr}\right)_e - \rho_e\left(\frac{dT}{dr}\right)_e}{\frac{\mu_c}{\mathcal{R}}\left(\frac{dP}{dr}\right)_c - \rho_c\left(\frac{dT}{dr}\right)_c}$$

$$= \frac{\frac{\mu_c}{\mathcal{R}}\cdot\frac{\mu_e}{\mu_c}\cdot\frac{\mu_e}{\mu_c}\left(\frac{dP}{dr}\right)_c - \frac{\mu_e}{\mu_c}\rho_c\left(\frac{\mu_e}{\mu_c}\right)^{1+n}\left(\frac{dT}{dr}\right)_c\frac{(1+X)_e}{(1+X)_c}(1+u)}{\frac{\mu_c}{\mathcal{R}}\left(\frac{dP}{dr}\right)_c - \rho_c\left(\frac{dT}{dr}\right)_c}$$

$$= \left(\frac{\mu_e}{\mu_c}\right)^2 \cdot \frac{\frac{\mu_c}{\mathscr{R}}\left(\frac{dP}{dr}\right)_c - (1+u)\left(\frac{1+X_e}{1+X_c}\right)\left(\frac{\mu_e}{\mu_c}\right)^n \rho_c \left(\frac{dT}{dr}\right)_c}{\frac{\mu_c}{\mathscr{R}}\left(\frac{dP}{dr}\right)_c - \rho_c\left(\frac{dT}{dr}\right)_c}$$

or

$$\frac{(d\rho/dr)_e}{(d\rho/dr)_c} = \left(\frac{\mu_e}{\mu_c}\right)^2 \frac{(\alpha - \gamma\beta)}{(\alpha - \beta)} \tag{147}$$

where

$$\gamma = (1+u)\left(\frac{1+X_e}{1+X_c}\right)\left(\frac{\mu_e}{\mu_c}\right)^n \tag{148}$$

(Note that the density gradients are negative.)

In a homogeneous star with no shell source, $X_e = X_c$, $\mu_e = \mu_c$ and $u = 0$; hence $\gamma = 1$ and by (147) the density gradient does not change at the 'core' boundary. However, when all the hydrogen in the core has been converted to helium, $X_c = 0$ and $\mu_c = 4/3$ (ionised helium).

In the star illustrated in figure 10, $\mu_e = 0.63$ and consequently

$$\left(\frac{1+X_e}{1+X_c}\right)\left(\frac{\mu_e}{\mu_c}\right)^n = 1.67\ (n = 0),\ \text{or}\ 0.79\ (n = 1)$$

If $u > 0.3$, $\gamma > 1$ even when $n = 1$, and $(\alpha - \gamma\beta)/(\alpha - \beta) < 1$. It follows from (147) that

$$\frac{(d\rho/dr)_e}{(d\rho/dr)_c} < \left(\frac{\mu_e}{\mu_c}\right)^2 = 0.22 \tag{149}$$

that is, the density gradient on the envelope side of the boundary is less than a quarter of the gradient on the core side: in comparison with a homogeneous star, the envelope must expand relative to the core by a factor of four or more. The core will have contracted somewhat, until the temperature at its surface has become high enough for nuclear synthesis to operate as it formerly did at the centre, but this is a relatively small change because the temperature at the core boundary was never far below that at the centre (see figure 6). Most of the effect appears as an actual expansion of the envelope (figure 11).

The expansion can be by a much larger factor than that calculated above if $u \gg 1$; but u cannot become infinite, because as the core contracts it releases gravitational energy and so L_c never becomes zero (see equation 144). Incidentally, since the envelope is expanded, it gains gravitational potential energy at the expense of the core. This

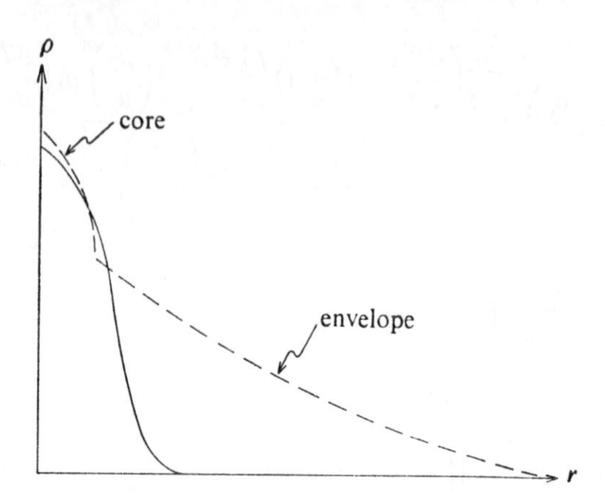

Figure 11. Contrasting the density distribution within a homogeneous main sequence star (solid line) with that in a red giant having a hydrogen-exhausted core (dashed line).

large expansion of the envelope turns the star into a red giant, since $T_e = L^{\frac{1}{4}} R^{-\frac{1}{2}}$ and L changes little.

The change in radius comes about through the need to maintain continuity of energy flux (hence the amount of expansion depends on the opacity law). If dT/dr and $d\rho/dr$ remained as for a homogeneous star, L would increase discontinuously across the boundary $r = r_c$ as a result of the discontinuous fall in density, since

$$L \propto r_c^2 T_c^3 (dT/dr)/\kappa\rho.$$

The temperature gradient dT/dr must fall to balance the fall in ρ: this the star achieves by expanding its envelope.

It may be noted that even if $u = 0$, the envelope will still expand when $n = 0$ (electron scattering) and may do so when $n = 1$, depending on the relative sizes of α and β (in fact solution of the equations of stellar structure (36) to (39) with the additional conditions (141) shows that it does). So a star becomes a red giant when it has exhausted the hydrogen in its core and before the shell source of nuclear energy has become operative; the shell source serves to accentuate the expansion of the envelope.

The contraction of the core and expansion of the envelope will proceed until the temperature has risen sufficiently to bring another nuclear reaction into operation. This may be the synthesis of helium from hydrogen in a shell source at the core surface, or the synthesis of heavier elements from helium at the centre of the core. It was noted in considering (108) and (109) that the probability of interpenetration of colliding nuclei decreases exponentially as the product of the atomic numbers of the nuclei; consequently fusion of helium nuclei requires a higher temperature to increase the rate and energy of collisions sufficiently to give a large enough rate of release of nuclear energy. It is not self-evident whether a hydrogen shell source or a helium core source will come into operation – it depends on whether the surface or centre of the core reaches the appropriate temperature first and this can only be determined by solution of the equations of stellar structure.

When $M_c \ll M_e$ (as in dwarf stars) the structure of the star is determined by M_e and the core readjusts to meet the interface conditions. When $M_c \gg M_e$ the structure is determined by the core and the envelope readjusts, but since it contains little mass it does not greatly alter the overall radius. When $M_c \approx M_e$ both readjust, and the change in radius is a maximum; this is the case for massive stars, since they have about half the mass in a convective core. Consequently when a main sequence star becomes a red giant, thc resulting change in radius increases with increasing stellar mass.

It turns out that as the core contracts, the change in density gradient in the envelope ($r > r_c$), required to maintain hydrostatic equilibrium and to conserve energy flow, becomes progressively larger than the factor $(\mu_e/\mu_c)^2$ and the overall radius of the star may increase by a factor of a hundred. Especially is this so when a shell source of energy occurs which would cause dT/dr to steepen further if the envelope did not expand more. Referring to (147), γ approaches α/β and does so more closely as u increases. When this happens, the envelope gains so much gravitational potential energy that the core has to contract quickly to release sufficient energy for this and for the luminosity as well, since from (65),

$$L = \frac{dE}{dt} = \frac{3\gamma-4}{3(\gamma-1)}\left(\frac{d\Omega(\text{core})}{dt} + \frac{d\Omega(\text{env})}{dt}\right)$$

and $d\Omega(\text{env})/dt$ is negative and nearly as large as $d\Omega(\text{core})/dt$. The evolution to the red giant state occurs quickly, leaving a gap devoid of stars between the main sequence and red giant sequence on the luminosity–temperature diagram. This is known as the Hertzsprung Gap.

3.2. *The Evolution of a Star of Mass 5M⊙.*

We can now examine in some detail the evolution of a star through the main sequence and red giant stages. This will be done with reference to the schematic L, T_e diagram in figure 12 which is based on the description by I. Iben in *Annual Review of Astronomy and Astrophysics*, 1967. In later chapters the pre-main sequence and post red giant stages will be examined.

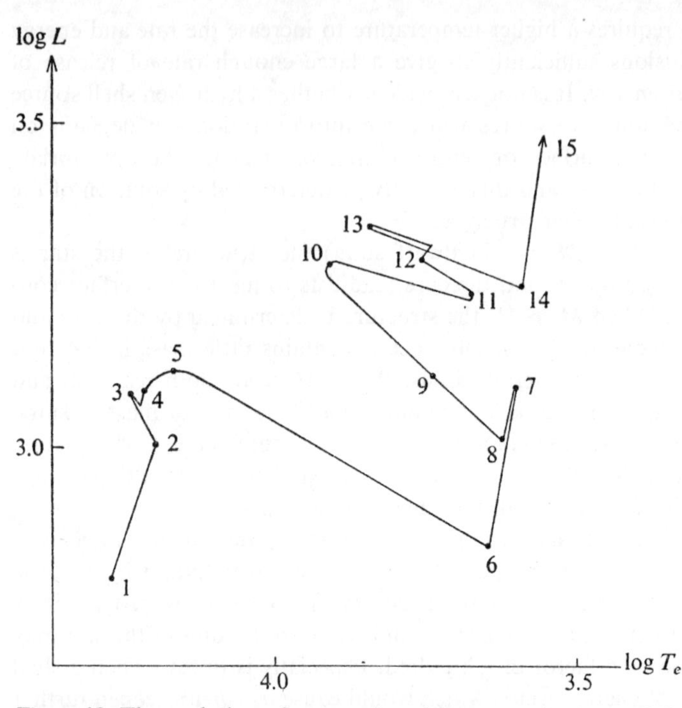

Figure 12. The evolution of a star of moderate mass in terms of the observable parameters of luminosity and surface temperature. The numbers identify stages described in the text.

The details of the evolution depend on the mass and chemical composition of the star; this account will relate to a star of five solar masses with a hydrogen abundance X of 71 per cent by mass, helium abundance Y of 27 per cent and an abundance Z of heavier elements 2 per cent. The relative abundances of the various heavy elements within that 2 per cent are assumed to be the same as in the Sun in the first instance. The evolution will be discussed in stages, the points where there are significant changes in the evolution being marked 1, 2, 3, etc. on figure 12.

Stage 1 *to* 2. The equilibrium between luminosity and energy generation by the CN cycle, described by figure 8, is reached at 1 on figure 12. The region in which a significant amount of energy is produced is close to the centre of the star because the reaction rate depends so steeply on temperature, and it occupies a smaller volume than the convective core. The products of the reactions are mixed by the convection through the volume of the core. However, as the CN cycle reactions proceed, not only does the abundance of helium increase at the expense of hydrogen but the abundances of N^{14} and

C^{12} change; also the abundance of N^{14} is substantially increased by the reaction.

$${}_8O^{16} + {}_1H^1 = {}_9F^{17} + \gamma$$
$${}_9F^{17} \rightarrow {}_8O^{17} + \beta^+ + \nu$$
$${}_8O^{17} + {}_1H^1 = {}_7N^{14} + {}_2H_e^4$$

Two effects now operate to change the luminosity progressively: the increasing abundance of N^{14} makes the CN cycle more efficient so that it generates enough energy at a slightly lower temperature, and the increasing molecular weight μ due to conversion of hydrogen to helium causes the luminosity to increase. The amounts of these changes can be estimated roughly as follows.

The opacity within much of the mass of the star will be due to electron scattering: this can be seen by comparing κ derived from (89) with that derived from (98) which shows that at a temperature 2×10^7 K electron scattering is about six times more effective than the bound-free transitions which dominate in the Kramer's formula.

It follows that the mass luminosity law as given by (16) is

$$L \propto \frac{\mu^4 M^3}{(1+X)}$$

Energy generation by the CN cycle is, from (134),

$$\bar{\varepsilon} M \propto Z(N^{14}) T_c^{26}$$

For equilibrium these must be equal, whence

$$T_c \propto Z(N^{14})^{-1/26} \tag{150}$$

Calculation of the rates of the reactions in the above cycle in the manner earlier carried out for the *pp* and CN cycles shows that the abundance of N^{14} increases by a factor of about ten. Consequently by (150) the change in T_c is $\Delta \log T_c = -0.04$. This is shown on figure 13.

The change in the composition of the core when all the hydrogen has been converted to helium is from $X = 0.71$, $Y = 0.27$, $Z = 0.02$, $\mu = 0.61$, to $X = 0$, $Y = 0.98$, $Z = 0.02$, $\mu = 1.34$. The core contains 14 per cent of the mass of the star and so averaged over the mass of the star, μ changes from 0.61 to 0.71 and $(1+X)$ from 1.7 to 1.6. Again using (16) quoted above, this gives $\Delta \log L(\mu) = +0.29$, and this is also shown on figure 13.

Averaging over the mass of the star in this crude manner is not entirely justified – a proper calculation requires solution of the equations of stellar structure for both conditions – but in fact the result is very close to that obtained by an accurate solution for the model considered (+0.26). This is not really surprising when one considers why μ^4 occurs in (16); it originates in the combination of the pressure derived from the equation of hydrostatic equilibrium with that given by the perfect gas law. It is therefore to be expected that the effect of local changes in μ will be weighted in proportion to the total mass of the star affected.

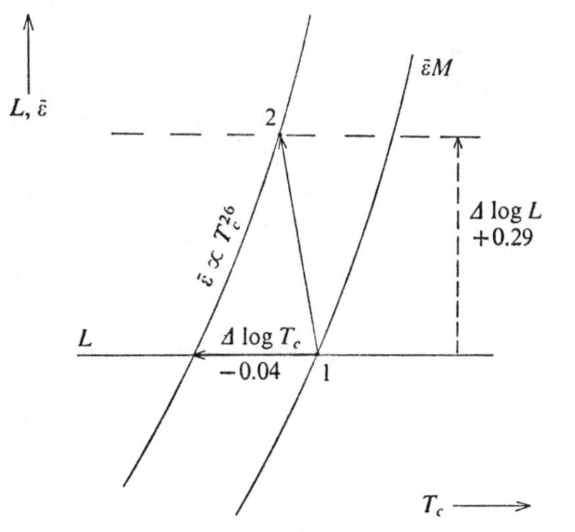

Figure 13. Changes in central temperature and energy production and outflow during evolution from 1 to 2 on figure 12.

Referring to figure 13, the equilibrium between L and $\bar{\varepsilon}M$ moves, during this main sequence evolution, from point 1 to 2. Since $L = R^2 T_e^4$ and $T_c \propto \mu M/R$ it follows that $T_e^4 \propto LT_c^2/\mu^2$. Inserting the values obtained above gives $\Delta \log T_e = +0.02$. Thus the surface temperature remains essentially constant, the evolutionary track on the L, T_e diagram moving almost vertically upwards, figure 12.

Although the mass of the convective core was quoted above as 14 per cent of the total mass, that was an average during this phase of evolution, because it does not stay constant. The condition for convection is from (27), that the temperature gradient must be steeper than the equality in $d \log T/d \log P \geqslant (\gamma - 1)/\gamma$. In a perfect gas where radiation pressure is much less than gas pressure, $\gamma = 5/3$ and the right-hand side takes the value 0.4.

Now from (8) and (15),

$$\frac{d \log T}{d \log P} \propto \frac{\kappa L}{R^4 T^4} \qquad (151)$$

In examining the value of this in the core, T can be written T_c, and R^4 refers to the star as a whole since it derives from the density and the equation of hydrostatic equilibrium. Note that $\Delta \log R = \Delta \log \mu - \Delta \log T_c$, where the change in μ is the average over the star. L is produced close to the centre, and κ is the value in the core since it affects the temperature gradient locally. During hydrogen fusion $\kappa = 0.2(1+X)$ has changed from 0.34 to 0.21 in the core. Inserting the changes in L, R and T_c derived above gives

$$\Delta \log\left(\frac{d\log T}{d\log P}\right) = -0{\cdot}22$$

that is, the temperature/pressure gradient has decreased by almost 70 per cent, a consequence primarily of the reduced opacity allowing the radiation through more freely. As a result, the condition for convection no longer applies except close to the energy-producing volume near the centre. The volume of the convective core, which initially embraced 20 per cent of the mass, now contains only 8 per cent.

The time taken to evolve from 1 to 2 on figure 12 is easily calculated. Of the total mass of the star of 10^{34} g, an average 14 per cent is in the convective core and thereby available to the nuclear reactions, and 71 per cent of this matter is hydrogen; thus the mass of hydrogen available is 10^{33} g, and synthesis to helium releases 6.3×10^{18} erg g^{-1} (section 2.10(a) and equation 124). This gives the total energy produced as 6.3×10^{51} erg. It is radiated away at a rate equal to the average luminosity between 1 and 2; solution of the equations of structure (36) to (39), or observations (equation 4 or figure 1), show the luminosity at 1 to be 2×10^{36} erg s^{-1}, and consequently the average from 1 to 2 is 3×10^{36} erg s^{-1}. At this rate, the time taken to radiate 6.3×10^{51} erg is 2.1×10^{15} s or 7×10^{7} yr.

Stage 2. To describe the star adequately, account must be taken of the inhomogeneity in composition between core and envelope. Using (147), in the present state $u=n=0$ and $\gamma=1.63$; hence the ratio of the density gradient on the envelope side of the core boundary to that on the core side has become

$$\frac{(d\rho/dr)_e}{(d\rho/dr)_c} < \left(\frac{\mu_e}{\mu_c}\right)^2 = 0.25$$

Most of the mass is, however, in the envelope and the core is forced to make most of the necessary adjustment; the overall increase in R is only by a factor of 1.7.

Stage 2 *to* 3. During this phase, the hydrogen abundance at the centre decreases to zero. ε(CN) declines to zero in the core and ε(gravitational) increases as the star tries to restore ε(CN) by contracting to raise T_c, leading to a slight overall contraction at a rate given by (139). This continues until the temperature T_i at the interface between core and envelope has risen high enough for hydrogen synthesis to begin in a shell.

Equation (139) gives for the rate of contraction $dR/dt = 0.2 \text{ cm s}^{-1}$; since the radius is now 5×10^{11} cm and only a modest contraction is required to raise T_i to the required value, this phase is very short, lasting only for some 10^4 yr.

Stage 3 *to* 4. The CN cycle now begins to operate in a shell around the hydrogen-exhausted core and increases until it meets all the energy needs of the star.

Because of the progressive shrinkage of the volume of the convective core during evolution from 1 to 2 there is not a sharp boundary between an all helium core and a hydrogen-rich envelope, but a transition zone in which the hydrogen abundance decreases inwards. The shell therefore begins fairly near to the centre where X is low but T higher, and extends outwards since X rises as T falls with increasing radial distance.

Since the core finally no longer produces energy, its heat content flows out as long as there is a temperature gradient in it; it is surrounded by an energy producing shell at temperature T_i and consequently the core cools until it becomes isothermal at that temperature. This readjustment occupies some 10^5 yr.

Stage 4 *to* 5. The shell source, containing about 5 per cent of the total mass, moves slowly outwards as it synthesises hydrogen to helium, leaving behind an ever-growing mass of helium in the core. The core radius contracts slowly to maintain the temperature at the shell adequate for the nuclear reactions, but the consequent rate of release of gravitational energy is small and the core remains essentially isothermal.

Stage 5 *to* 6. The mass in the isothermal core cannot continue to grow until it embraces the whole mass of the star. Two effects prevent it from containing even a major fraction of the mass.

First, the pressure in the core still has to support the weight of the overlying layers. Since the temperature is uniform, the whole of the increase in pressure to the centre has to be met by an increase in the density. The central density therefore has to become very high, and the core of the star has to contract to bring this about. There is no 'static' solution with a large fraction of the mass in the core – the contraction increases the self-gravitation and the required pressure increases still further, and so on; the equation of hydrostatic equilibrium cannot be met even though the central density were to increase to infinity. The gravitational energy released by the contraction averts this collapse because it ultimately results in a temperature gradient developing to carry the energy outwards and to increase the pressure more towards the centre.

Secondly, there is a limit to the temperature gradient which the envelope can bear. As the core contracts to increase its central density the envelope expands, as we found in section 3.1, to retain hydrostatic equilibrium and continuity of energy flow. As the mass of the isothermal core grows larger these become more difficult to meet, and the contraction of the core and expansion of the envelope occur with increasing speed. The core releases more and more gravitational energy, the temperature gradient steepening to carrying it outwards. The envelope expands to an enormous radius, taking the star rapidly across the L, T_e diagram.

As a result of the increase in radius, the opacity due to bound-free transitions increases substantially. It will be recalled that it depends on density and temperature as $\kappa \propto \rho T^{-3.5}$. If a star expands without changing its structure, that is to say it just 'scales up' in size, the relationship between the changes of ρ and of T at any relative distance from the centre is easily derived, since as R changes $T \propto \mu M/R$ and $\rho \propto M/R^3$ at any point r/R, the constants of proportionality depending on the mass distribution (the structure) and on the value of r/R; thus $\rho \propto T^3$ (note that this does not give the dependence of ρ on T as r/R is varied, only the dependence of ρ on T at any given r/R as R is varied). Consequently at any given r/R, $\kappa \propto T^{-0.5} \propto R^{0.5}$ as R is varied.

It is sufficient that the mass distribution exterior to r does not change as R changes. This is nearly true for the atmosphere of the red giant as it expands nearly twenty-fold; the opacity at any point in it due to bound-free transitions increases by a factor four.

The condition for convection is, from (27), $d \log T/d \log P > (\gamma-1)/\gamma$. But

$$\frac{d \log T}{d \log P} = \frac{P}{T}\frac{dT}{dr}\Big/\frac{dP}{dr}$$

$$= -\frac{P}{T}\frac{\kappa\rho L}{r^2 T^3}\Big/\frac{\rho G M}{r^2}$$

using (31) and (34). At any given r/R, this becomes

$$\frac{d \log T}{d \log P} \propto \mu^{-4} M^{-3} L \kappa$$

and since outside the energy producing region L is constant, with constant total mass and molecular weight this gives,

$$\frac{d \log T}{d \log P} \propto \kappa \tag{152}$$

as R changes. The fourfold increase in κ resulting from the expansion of the envelope causes the criterion (27) to be met over a large extent of the atmosphere, which therefore becomes convective.

As convection spreads down to regions where nuclear reactions have changed the relative abundances of the elements – for example the conversion of lithium to He^3 which takes place at a temperature of about a million degrees–the surface abundances of the product elements change.

The time for transition from the region of the main sequence to that of the red giants is short, only some 8×10^5 yr, and consequently few stars are observed in this transition stage.

On reaching point 6, the change in the interior structure has resulted in a fall in the luminosity; note that this is due to the changes in the terms in the relation (34). This, together with the increase in radius, causes the effective temperature of the surface of the star to fall to 4 500 K.

Stage 6 *to* 7. Energy is now transported from the interior almost entirely by convection. Since the energy must be radiated from the star it must, of course, be transferred to a radiative atmosphere. This atmosphere is a blanket, the opacity of which controls the rate of emission of energy from the star. The atmosphere extends outwards from the hydrogen ionisation convection zone (see p. 15) where $T \approx 10000$ K, to the photosphere at 4500 K. The opacity in the atmosphere is due largely to dissociation of the negative hydrogen ion, but the abundance of this ion decreases with decreasing temperature in this temperature range because the electrons come from ionisation of metals, and therefore the opacity decreases outwards in the atmosphere: the opposite to Kramer's opacity. Now if the envelope expands a little and cools, the opacity falls and radiation escapes from regions deeper down at higher temperature, causing an increase in the luminosity. This more rapid escape of heat causes the stellar envelope to contract and heat up, increasing the opacity until the radiation is once more escaping from a layer at $T_e \sim 4500$ K. Thus the temperature stays close to 4 500 K at all luminosities in this extended red giant condition. The evolutionary track on the L, T_e diagram, figure 12, is along an almost vertical line.

Convection now extends inwards to the outer part of the original convective core where the CN cycle has converted most of the original C^{12} to N^{14} (see section 2.13), and the surface abundances change accordingly. At the centre, the temperature is high enough for the operation of the cycle

$$_7N^{14} + _2He^4 = _9F^{18} + \gamma \rightarrow _8O^{18} + \beta^+ + \nu$$

which converts almost all the N^{14} into O^{18}.

This phase lasts for about 5×10^5 yr, the luminosity and central temperature increasing progressively.

Stage 7. The central temperature reaches 1.5×10^8 K, and synthesis of helium to carbon, the so-called triple-α reaction, begins:

$$\left.\begin{array}{ll} \text{(i)} & {}_2\text{He}^4 + {}_2\text{He}^4 + 95 \text{ KeV} = {}_4\text{Be}^8 \\ & {}_4\text{Be}^8 \rightarrow {}_2\text{He}^4 + {}_2\text{He}^4 + \gamma \text{ (lifetime } 10^{-14} \text{ s)} \\ & \text{Equilibrium abundance of } {}_4\text{Be}^8 \text{ is } \sim 10^{-10} n_{\text{He}} \\ \text{(ii)} & {}_4\text{Be}^8 + {}_2\text{He}^4 = {}_6\text{C}^{12} + \gamma (7.4 \text{ MeV}) \end{array}\right\} \quad (153)$$

This last reaction has a strong resonance at a collision energy $E_r = 310$ keV. Consequently the rate of the reaction is governed by the rate of collisions in the resonance width dE_r at E_r.

The rate of the 3α cycle is then calculated as follows. From (114),

$$\frac{dN}{dt} = \frac{8\pi N_1 N_2}{(2\pi kT)^{\frac{3}{2}}} \frac{\sigma_r}{m^{\frac{1}{2}}} e^{-E_r/kT} E_r \, dE_r \qquad (154)$$

where m is the reduced mass, $N_1 = N(\text{He}^4)$ and $N_2 = N(\text{Be}^8)$. The 'Saha' equation gives for the equilibrium between $N(\text{He}^4)$ and $N(\text{Be}^8)$,

$$N(\text{Be}^8) = (N(\text{He}^4))^2 \frac{h^3}{(2\pi m(\text{He}^4)kT)^{3/2}} \frac{g(\text{Be}^8)}{2g(\text{He}^4)} e^{\chi_{\text{Be}}/kT} \quad (155)$$

where the g are partition functions and $\chi_{\text{Be}} = -95$ keV is the energy given up by the disintegration of Be^8. Now

$$m = \frac{m(\text{Be}^8)m(\text{He}^4)}{m(\text{Be}^8)+m(\text{He}^4)} = \frac{2(m(\text{He}^4))^2}{3m(\text{He}^4)} = \tfrac{2}{3}m(\text{He}^4)$$

therefore $m(\text{He}^4) = \frac{3}{2}m$. From (115),

$$\sigma_r = \frac{\pi \hbar^2}{2mE_r} = \frac{h^2}{8\pi m E_r}$$

where the reduced mass is used because the cross-section is derived for the relative momentum. Also, from (116),

$$dE_r = \Gamma_r \hbar = \frac{\Gamma_r h}{2\pi}$$

where Γ_r is the reciprocal mean lifetime. Consequently, substituting in (154) for N_1, N_2, $m(\text{He}^4)$, σ_r and dE_r, and using (155), the reaction rate becomes after some straightforward reduction

$$\frac{dN}{dt} = \frac{2^{3/2}}{3^{3/2}4\pi} \cdot \frac{\Gamma h^6}{(2\pi mkT)^3} \left(\frac{Y\rho}{4m_{\text{H}}}\right)^3 \frac{g(\text{Be}^8)}{g(\text{He}^4)} e^{-(E_r - \chi_{\text{Be}})/kT} \quad (156)$$

where Y is the helium abundance by mass.

Each of these resonant collisions leads to a transmutation, and the energy given up per transmutation is $Q = 7.305$ MeV. The rate of energy generation per unit mass is then given by

$$\varepsilon(3\alpha)=\frac{Q}{\rho}\frac{dN}{dt} \qquad (157)$$

The power law approximation to this, calculated after the manner of (125) and (126) is

$$\varepsilon(3\alpha)=1.6\times10^{-8}\rho^2y^3T^{28}\ \text{erg g}^{-1}\ \text{s}^{-1} \qquad (158)$$

in the region of $T=1.5\times10^8$ K.

At point 7 on figure 12, ignition of the 3α process causes a sudden overproduction of energy at the star's centre as a result of the high temperature dependence of (158). The energy transport mechanism cannot handle this large additional energy output, the centre of the star heats up and expands and the luminosity falls. We can examine how this comes about by reference to (61), (62) and (64). From them, remembering that E is the total energy and U is the thermal energy: ΔE has the same sign as ΔR; ΔU has opposite sign to ΔR; and $\Delta\Omega$ has the same sign as ΔR.

Consequently when overproduction of energy heats the core of the star, and thus by increasing the pressure $P=\rho\mathcal{R}T/\mu$ causes it to expand: Ω(core) increases, the heat does work in expanding the gas; U(core) decreases, and since $U=C_VT$, T falls; and E(core) increases, therefore the energy being lost, L, must decrease. So, supplying additional energy – heating the star by an additional energy source other than gravitation – causes the star to *cool*!

What is happening is that when the extra heat is supplied, and the core of the star expands to a new stable configuration, more energy than the extra heat supplied is used in expanding the star to that stable configuration; this additional energy is withdrawn from the thermal content of the gas and goes, via work done, into gravitational energy. The effect is therefore not so remarkable as appears at first sight, and indeed if it was not present the star could be unstable – that is, if increased nuclear energy production caused the temperature to rise rather than to fall. The very existence of the star is a consequence of the effect.

On a P, V diagram the effect appears as on figure 14. The change in thermal gas pressure with change in temperature or volume is given by the gas law $P(\text{thermal})\propto TV^{-1}$, whereas the pressure required to balance self-gravitation is, by the proportionality (8), $P(\text{gravity})\propto M^2/R^4\propto V^{-4/3}$ for given M.

Beginning in hydrostatic equilibrium at P_1, T_1 on figure 14, add heat dQ to the gas; $dQ=C_VdT+PdV$. Expand isothermally to P_3, T_1; P(thermal) changes as V^{-1}; but P(gravity) has changed as $V^{-4/3}$ and has fallen further; to achieve pressure equilibrium, cool at constant volume (that is, doing no work) to P_2, T_2, and hydrostatic equilibrium is restored.

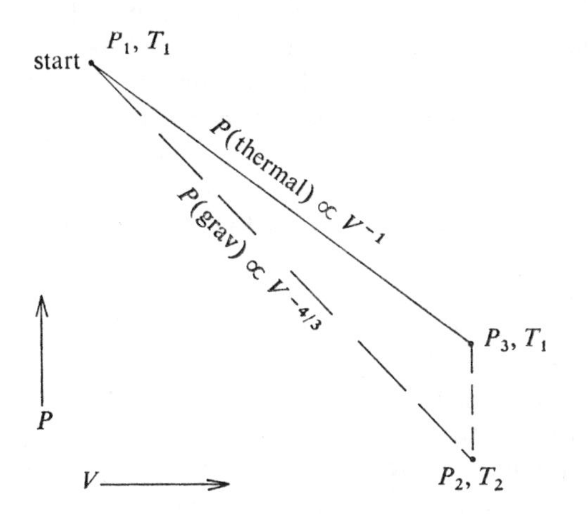

Figure 14. The P, V diagram showing that the thermodynamic behaviour of the star causes it to cool when heat is added due to overproduction of energy. Thermal energy is removed and put into gravitational potential energy. Without this property stars would not be stable.

The rapid expansion of the core and consequent reduction in luminosity, due to the sudden onset of the 3α process (often called the helium flash), causes a spike on the evolutionary track, figure 12.

Stage 7 *to* 8. Ignition of the 3α process has once again provided a central energy source in the convective core. The energy produced by the shell source must now fall, so that $L(\text{shell}) = L - L(\text{core})$; a readjustment of the density distribution occurs to reduce the total energy production to that necessary to balance the luminosity. This change in structure itself causes the luminosity to decrease somewhat. Furthermore, reduction in the strength of the shell source leads, as shown in section 3.1, to a contraction of the envelope, so that despite the fall in the luminosity the surface effective temperature increases slightly.

With the energy of helium synthesis at its disposal, the star remains in this part of its evolution rather longer than in the other stages since leaving the main sequence: some 6×10^6 yr.

Stage 8 *to* 10. The energy produced by the shell continues to decline relative to that produced in the core so that the envelope continues to contract. As shown in the discussion leading up to (152), this causes $d \log T / d \log P$ increasingly to fall below the initial value for convection, and so convection dies out throughout a large part of the envelope.

The nuclear energy source, however, proves inadequate and, as the star contracts to raise its central temperature, release of gravitational energy contributes substantially to the heat supply. The opacity throughout much of the envelope is still due to bound-free transitions, and since, from (14), for a given mass distribution $L \propto \mu^{7.5} M^{5.5} R^{-0.5}$, L increases as R decreases. Also, since $L \propto R^2 T_e^4$ it follows that

$$\Delta \log L = \tfrac{4}{5} \Delta \log T_e \qquad (159)$$

and this is the slope of the evolutionary track on the L, T_e diagram, figure 12, from point 8 to point 10. The time spent on this track is 10^7 yr.

Stage 10 *to* 11. At 10 the central temperature has risen high enough for the 3α cycle to supply the immediate energy needs alone; $L = M\bar{\varepsilon}(3\alpha)$. As most of the helium in the convective core becomes converted to carbon, the core contracts until the temperature at its surface is high enough for the 3α cycle to operate in a shell. The molecular weight $\mu_c \to 2$. Increasing μ_c/μ_e and $L(\text{shell})/L$ cause renewed expansion of the envelope. The evolution is analogous to that in stages 4, 5 and 6, but with the 3α cycle supplying energy instead of the CN cycle. This is a fairly rapid phase of evolution, lasting only 10^6 yr.

Stage 11 *to* 12. At 11, convection has ceased in the core with the exhaustion of helium there, and the structure adjusts to a radiative core. With the ending of energy production and convection the temperature gradient decreases and the core expands. The work done in expanding the core and thus increasing the gravitational potential energy is at the expense of the thermal energy. The core, and the interface between core and envelope, cool; $\bar{\varepsilon}M$ falls below L and the whole star contracts to release gravitational energy until the shell temperature has risen high enough for nuclear energy production to balance the outflow of radiation: a similar readjustment occurred from 2 to 3. This is a short-lived affair, taking barely 10^5 yr.

Stage 12 *to* 13. All this time, the CN cycle has been operating in a shell source further from the centre than the 3α cycle, but now the 3α cycle is producing much more energy and the weighted 'mean position' of the shell source effectively moves inward. Thus the 'core' effectively becomes smaller relative to the envelope, and the radius decreases (see p. 55).

Stage 13 *to* 14. The CN shell source is largely exhausted, the hydrogen having all been synthesised to helium where the temperature is high enough for the cycle to operate. All the energy comes from the synthesis of helium to carbon in the 3α shell source. As the shell source consumes the helium and moves progressively outwards, increasing the mass of the carbon core, the envelope expands.

Evolution is analogous to stage 5 to 6 but with a C^{12} core surrounded by a 3α shell source. The expansion of the envelope causes the opacity to increase, and a convection zone appears and deepens (see stage 5 to 6). Convection reaches the helium previously synthesised by the CN shell source and mixes it to the surface.

The core contracts to relativistic degeneracy, the equation of state in the core now being given by (81) instead of by the perfect gas equation.

Stage 14 *to* 15. The immediate course of evolution depends on whether or not large neutrino losses occur – still a matter of uncertainty. If they do occur, they carry large amounts of energy out of the core without heating the envelope in their passage through it.

a) Assuming neutrino loss. The centre of the core cools as it contracts to relativistic degeneracy as a result of the energy carried away by neutrinos. The outer part of the core is not emitting neutrinos and heats up as it contracts, maintaining the shell temperature required for the operation of the 3α cycle. This shell has become very thin because of the steep temperature gradient and the high sensitivity of the rate of the cycle to temperature, $\varepsilon(3\alpha) \propto T^{28}$. As a consequence it becomes thermally unstable, any slight overheating causing a rapid overproduction of energy before the star can expand and cool. This is because the thermal inertia of the volume occupied by the shell is much less than the mass inertia of the region plus that of the overlying layers (which have to be lifted if the shell is to expand). The course of events is illustrated on figure 15. The temperature and hence the pressure begin to rise.

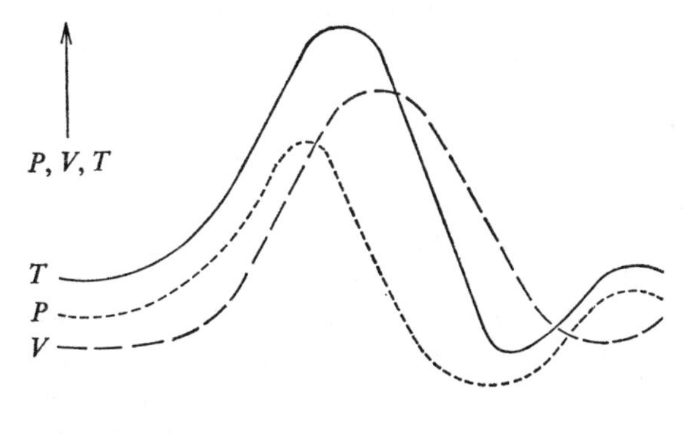

Figure 15. Rapidly changing conditions in the star lead to instability, because the mass inertia is so much greater than the thermal inertia that it lags and the forces illustrated on figure 2 no longer balance. Pulsation results.

The volume expansion lags at first, being slow because of the need to accelerate the overlying mass, but the large inertia of this mass causes it to overshoot. This over-expansion causes the temperature and pressure to fall below their initial values, and the material, no longer fully pressure-supported against the gravitational attraction of the mass interior to it, falls back inwards to start the cycle again. The envelope pulses with increasing amplitude and a period of several thousand years; there is the possibility that a considerable amount of the stellar mass may ultimately be 'blown off'.

b) Assuming no neutrino loss. The centre of the core is degenerate. The equations of state for a degenerate gas, (71) and (72) or (77) and (81), show that the pressure depends only on the density and not on the temperature. These equations were derived on the assumption that all the quantum cells $(\Delta q)^3 = (\hbar/p)^3$ are filled by electrons; that is, that $n_e = n_p = 1/(\Delta q)^3$. However, if energy is added to the gas, the momentum per electron, p, increases since $p = (2mE)^{\frac{1}{2}}$; then Δq falls and $1/(\Delta q)^3 > n_e$, so that some of the phase cells are empty. The degeneracy has been partly removed and the pressure becomes to some extent dependent on the temperature, causing the gas to expand as heat is added.

Heat is added in two ways; first by the release of gravitational energy as the core contracts then, when the central temperature has become high enough, by nuclear reactions involving C^{12}. The removal of degeneracy and the C^{12} reaction at the centre cause a convective core to be established. The density and temperature distributions in the star readjust accordingly, the temperature in the region of the 3α shell falls and $\varepsilon(3\alpha) \rightarrow 0$; the temperature at the hydrogen–helium interface rises and a CN cycle shell is formed.

With or without neutrino loss, this phase of the evolution is brief, only about 10^4 yr.

Thereafter, the course of evolution is uncertain. The observational data indicates that the stars spend most of their remaining lives, before becoming white dwarfs, on a horizontal branch on the L, T_e diagram, figure 12, but it is not known in which direction evolution takes them along it. Some have argued for mixing, others for mass loss, to bring the stars from the top of the red giant branch to the hotter end of the horizontal branch. Although this phase is so uncertainly known, the subsequent fate of such a star is much more certain, if only because exhaustion of the nuclear fuel supplies makes it easier to determine what happens as the star continues to radiate energy with only gravitation to draw upon. It becomes a white dwarf.

In this chapter and in the preceding one, the structure and evolution of a star has been considered from the moment it first reaches equilibrium on the main sequence – where it spends the greater part of its life – to the stage where it has passed through the red giant phase and is running out of fuel. In the following two chapters we shall examine how it arrived on the main sequence in the first instance, and then what happens to it as it runs out of nuclear fuel and becomes a white dwarf.

4. Pre-Main Sequence Stars.

It is generally accepted that stars form by the break-up of massive interstellar clouds into fragments of stellar mass, and the contraction of those fragments to stellar densities. The processes by which protostars form are outside the scope of this book, but once a cloud fragment has become so dense that it will almost certainly continue to contract to become a star it has entered the realm of stellar evolution; from that stage to the main sequence is the subject of this short chapter.

The condition for contraction of a cloud of solar mass is given by the Virial Theorem (47),

$$\frac{3\mathscr{R}T}{\mu} < \frac{3}{2}\frac{GM}{R} \tag{160}$$

which is

$$\rho > \frac{3}{4\pi}\left(\frac{2\mathscr{R}}{\mu G}\right)^3 \frac{T^3}{M^2} \tag{161}$$

Investigations of dense interstellar clouds indicate that their temperatures are about 20 K, and with $M = 2 \times 10^{33}$ g this gives $\rho > 3 \times 10^{-18}$ g cm^{-3} as the density at which contraction begins for a cloud of solar mass. The time scale for contraction to high density at free fall rate is, from (7),

$$t \sim (6\pi\rho G)^{-\frac{1}{2}} \tag{162}$$

which for the density given is $\leqslant 2 \times 10^4$ yr. By astronomical time scales the pre-main sequence evolution is therefore a fairly short-lived phenomenon. Few stars are likely to be observed in this phase, although they do appear to be represented by rather rare objects such as some of the infra-red emitters.

At stellar densities and temperatures, the atoms are excited to high energy levels which are closely spaced, many absorbing transitions thus being available to capture the high energy photons passing through the material (section 2.6). Stellar material thus has high opacity. In the interstellar clouds a contrary situation exists: the clouds have low densities and temperatures; the atoms are in low energy states which are widely spaced and so cannot capture the weak photons emitted by the low-temperature gas, which is therefore highly transparent. The gas throughout a fragment is consequently exposed to the same radiation field and takes up a common temperature.

In equilibrium, before contraction begins, the pressure in such an isothermal cloud fragment must increase towards the centre to support the weight of the overlying layers, and consequently the density increases inwards similarly. It follows from (160) that when contraction gets under way it proceeds more quickly towards the centre, which therefore becomes relatively even more dense than the outer part. A very dense core forms rapidly. Throughout this time the temperature has been about 20 K, the hydrogen is molecular and most of the opacity is due to dust grains. When the core becomes opaque the energy released by gravitational contraction can no longer escape and it heats the gas. The H_2 molecules, now dissociated by collisions between the molecules, absorb the binding energy of 4.2 eV per dissociation and rapidly cool the gas, causing a further rather sudden collapse.

The optical depth due to dust grains in a cloud of radius r is

$$\tau = \pi a^2 Q n_g r \tag{163}$$

where $\pi a^2 Q$ is the optical cross section and n_g the number density of grains. The optical cross section is not well known at the wavelengths of interest, given by Wien's Law

$$\lambda_m T = 0.29 \text{ cm deg} \tag{164}$$

When $T = 20$ K, $\lambda_m = 0.14$ mm; estimates indicate that $\pi a^2 Q \approx 10^{-16}$ cm^2 at this wavelength. Since $n_g/n_{\mathrm{H}} \simeq 10^{-12}$, it follows from (163) that for a cloud of mass $M_\odot$ the optical depth becomes unity when the density rises to 10^{-10} g cm^{-3}.

The gravitational energy of this pre-star is

$$\Omega = \frac{3}{5}\frac{GM^2}{R} = 2 \times 10^{48} \rho^{1/3} \left(\frac{M}{M_\odot}\right)^{5/3} \tag{165}$$

which for $M = M_\odot$ at $\rho = 10^{-10}$ g cm^{-3} is

$$\Omega = 4.3 \times 10^{45} \text{ erg} \tag{166}$$

The heat is now largely trapped inside the material by the increasing opacity, and the gravitational energy being released by the continuing collapse heats the gas. The thermal energy is

$$U = \frac{3}{2}\frac{\mathscr{R}}{\mu} TM \tag{167}$$

To heat the gas to 10^4 K therefore requires 2.5×10^{45} erg: it follows from (165) and (166) that this will be released by a further contraction of the cloud to $\rho = 4 \times 10^{-10}$ g cm^{-3}; since some of the heat will be radiated away and some of the energy will go into dynamical motions the increase in density will be rather more than this. When the temperature reaches 10^4 K, collisions between the hydrogen atoms become violent enough to ionise them; the ionisation potential is 13.6 eV and the energy required to ionise all the gas in the star is

$$E(\mathrm{I}) = 2.6 \times 10^{46} \left(\frac{M}{M\odot} \right) \mathrm{erg} \qquad (168)$$

Again using (167), an increase in density from 4×10^{-10} g cm^{-3} to 4.5×10^{-8} g cm^{-3} will release the required gravitational energy. At this stage, the radius of the pre-star is $270R\odot$.

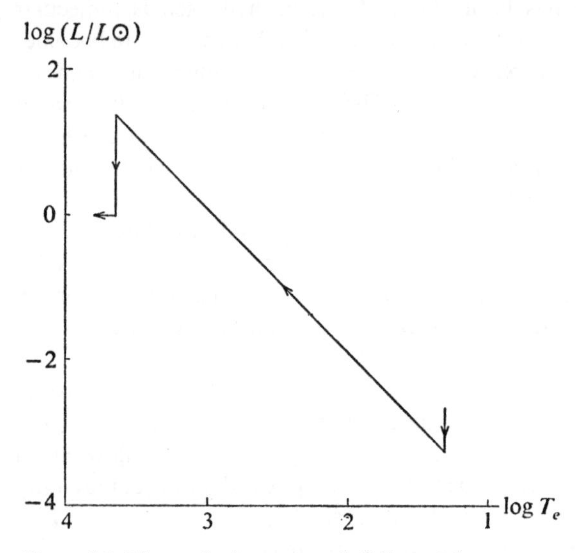

Figure 16. The evolutionary track followed by a proto-star contracting towards the main sequence.

As contraction continues and the surface temperature of the star rises, a state is reached at which the structure becomes similar to that of the red giant described in stages 6 to 7 of section 3.2; the surface temperature is ~ 4500 K, heat is transported outwards largely by convection, and the evolutionary track is almost vertical on the luminosity–temperature diagram. Convection rapidly transports the heat that had been trapped by the high opacity, the radius and hence the luminosity is high but falls as contraction continues (the only substantial source of energy still being gravity), and the central density increases until the relative density distribution is similar to that of a main sequence star. Contraction to the main sequence then proceeds homologously along a line given by (19) at constant mass and composition, *viz.*

$$L \propto R^{-0.5} \qquad (169)$$

That is, since $L \propto R^2 T_e^4$,

$$L \propto T_e^{0.8} \qquad (170)$$

The sequence of events described above are gathered together on figure 16. At the start of the contraction after fragmentation, the optical depth is low, $\sim 5 \times 10^{-5}$, and the amount of radiation emitted is thus very low despite the large radius. When the optical depth becomes of order unity at a density 10^{-10} g cm^{-3}, the radius is $2000R_\odot$ and, with the temperature still 20 K, the luminosity is very low, only $5 \times 10^{-4} L_\odot$. As contraction proceeds the interior heats up and the increasing gas pressure slows down the contraction. Under quasi-hydrostatic equilibrium the temperature then varies as given by (11),

$$T \propto R^{-1} \qquad (171)$$

and as contraction now proceeds more homologously,

$$L \propto R^2 T_e^4 \sim T_e^2 \qquad (172)$$

Thus the luminosity rises from $5 \times 10^{-4} L_\odot$ at $T_e = 20$ K to $25L_\odot$ at $T = 4500$ K when the convection track is reached, with a step on the way due to hydrogen ionisation. Then a rapid fall in luminosity is followed by the slow approach to the main sequence, finally halted on the main sequence by the onset of nuclear energy generation.

5. White Dwarfs, Neutron Stars and Pulsars. 5.1. 5.2. *Neutron Stars.* **5.3.** *Pulsars.*

5.1.

As a star runs through the various stages of evolution on the main sequence and then as a red giant, it draws on one nuclear fuel after another; hydrogen to build helium, helium to build carbon, carbon to build magnesium. The star always draws on its energy supplies to make energy production balance luminosity, and the luminosity is determined not by the fuel available but by the structure and mass of the star. The fuel enters into the equation only in that, by drawing on gravitational energy in the absence of any other, the star steadily contracts and, since $T_c \propto \mu M/R$, the central temperature increases until nuclear synthesis proceeds quickly enough to replenish the energy radiated by the star with that central temperature. When that particular nuclear fuel is exhausted, gravity is again drawn upon and the central temperature begins its upward climb until a new nuclear fuel becomes available at the higher temperature; and so on. In between periods that are almost static because each type of nuclear energy is generated at an almost constant central temperature, the star contracts slowly to higher and higher central temperatures and densities. Sooner or later it runs out of nuclear fuel, and also, in time, the central density becomes so high as to become degenerate (see section 2.5).

In the case of the most massive stars, this course of evolution ends in explosive instability, and that event will be dealt with in chapter 6; but the explosion probably leaves behind a degenerate core.

Other rather less massive stars may shed mass less violently; they and the more massive ones must shed mass because as we shall see there is an upper limit to the mass of a star which can become wholly degenerate, and no other course of evolution is open to it.

Stars of low mass become degenerate actually on the main sequence. They draw their energy from the *pp* cycle at a temperature which is approximately constant at about 10^7 K. Therefore $M/R \propto T/\mu$ is approximately constant with varying mass, and since $\rho_c \propto M/R^3$ then $\rho_c \propto M^{-2}$. Very low mass stars therefore have very high central densities.

In both the cores of red giants and in main sequence dwarfs, then, the density is so high that the gas becomes degenerate and the pressure is determined almost entirely by the density; thus, in non-relativistic degeneracy, equation (77),

$$P = \left(\frac{k_d}{\mu_e^{5/3}}\right)\rho^{5/3}$$

where $k_d = (3/8\pi)^{2/3}h^2/5m_e m_H^{5/3}$, and in relativistic degeneracy, equation (81),

$$P = \left(\frac{k_r}{\mu_e^{4/3}}\right)\rho^{4/3}$$

where $k_r = (1/8)(3/\pi)^{1/3}\ (hc/m_H^{4/3})$. The relationships are of polytropic form since the structure of a polytrope of index n is defined by the equations of hydrostatic equilibrium and continuity of mass together with a power law relationship between temperature and density, *viz.*

$$\left.\begin{aligned} &\rho \propto T^n, \text{ hence} \\ &P \propto \rho T \propto T^{n+1} = K\rho^{(n+1)/n} \end{aligned}\right\} \quad (173)$$

These relationships replace (38) and (39); they take no account of opacity or continuity of energy flow and may or may not represent a hot self-gravitating sphere of gas in equilibrium. For example, it has already been shown in section 2.2(c) that in convective equilibrium the pressure and temperature are related (equation 26) by $P \propto T^{\gamma/(\gamma-1)}$ where in a perfect gas $\gamma = 5/3$; hence $P \propto T^{5/2}$, and consequently $\rho \propto T^{3/2}$, a polytrope of index 3/2. Therefore the convective core of a star is a polytrope of index 3/2.

From the theory of polytropes,

$$M = 4\pi\left(\frac{(n+1)K}{4\pi G}\right)^{3/2} \rho_c^{(3-n)/2n}\left(-z^2\frac{du}{dz}\right)_{u=0} \quad (174)$$

where z and u are dimensionless variables, z being proportional to distance from the centre and u to temperature; consequently u^{n+1} is proportional to pressure and u^n to density.

The equation of state for non-relativistic degeneracy (77) therefore defines a polytrope where, by comparison with (173), $(n+1)/n = 5/3$; hence $n = 3/2$ and $K_d = k_d/\mu_e^{5/3} = 1.0 \times 10^{13}\mu_e^{5/3}$. Thus a non-relativistically degenerate star is a polytrope of index 3/2.

The mass of this polytrope is, by (174)

$$M = 5{\cdot}5 \times 10^{30}\mu_e^{-5/2}\rho_c^{\frac{1}{2}}\ \text{g} = 2.75 \times 10^{-3}\mu_e^{-5/2}\rho_c^{\frac{1}{2}}M_\odot \quad (175)$$

Note that $\rho_c \propto M^2$, contrary to the non-degenerate case noted above where $\rho_c \propto M^{-2}$.

Again from the theory of polytropes,

$$\frac{\rho_c}{\bar{\rho}} = \frac{-z}{3\,du/dz} \tag{176}$$

When $n = 3/2$, $\rho_c/\bar{\rho} = 6$. Consequently since $\bar{\rho} = 3M/4\pi R^3$, (175) can be written

$$R = 0.02\mu_e^{-5/3}M^{-1/3} \tag{177}$$

where R and M are in solar units.

It follows that if more mass is added to a degenerate star or to a degenerate core of a star, its radius decreases and its density increases, until it becomes relativistically degenerate; then from (81) and (173) it is represented by a polytrope for which $(n+1)/n = 4/3$, hence $n = 3$ and $K_r = k_r/\mu_e^{4/3}$. From (174), when $n = 3$, M is independent of ρ_c and is given by

$$M = 5.75\mu_c^{-2}M_\odot \tag{178}$$

(addition of radiation pressure only has the effect of changing the value of the numerical constant to 6.65).

Stars with masses less than this do not have densities high enough to be relativistically degenerate, their densities being given by (175); stars with greater masses would have too high a self-gravitation to be supported by the pressure represented by (81). Thus (178) gives an upper limit to the mass of a completely degenerate star, and to the mass of a degenerate core of a red giant. In practice of course, there would be a transition between non-relativistic and completely relativistic degeneracy, represented by polytropes with indices $3/2 \leqslant n \leqslant 3$.

If the evolution of a red giant star upwards of stage 14 on figure 12 continues to add mass to a degenerate fuel-exhausted core, then when the core mass reaches the limit set by (178) the star becomes unstable: it must eject the excess mass. It is possible that this ejection of the outer mass of a red giant is the origin of planetary nebulae.

When a star of mass $M \leqslant 5.75\mu_e^{-2}$ has contracted to the central density given by (175) it cannot contract any more and consequently it can neither heat up by releasing gravitational energy, nor reach a higher temperature to draw on more nuclear energy sources; it can only radiate its heat content and cool. The relation between luminosity and surface effective temperature is then given by $L = 9 \times 10^{-16}R^2T_e^4$ with L and R in solar units; or, using (177)

$$L = 1.44 \times 10^{-18}\mu_e^{-10/3}M^{-2/3}T_e^4 \tag{179}$$

where M is in solar masses. This relationship is illustrated on figure 17, where it coincides with the observed parameters of white dwarf stars and from which it is concluded that they are degenerate stars.

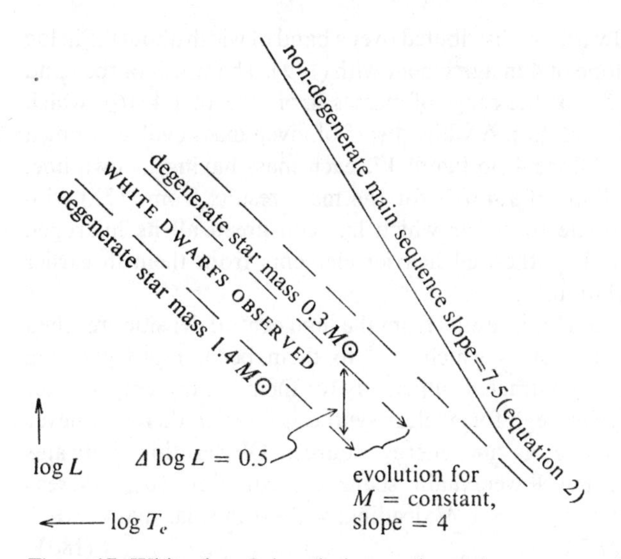

Figure 17. White dwarfs in relation to the main sequence, showing that the former are accounted for by degenerate masses.

Now observations show the lower part of the main sequence to have a slope of 7.5 on the log L–log T_e diagram, equation (2). Also, from (19) we have $L \propto R^5$ for constant $T_c \propto \mu M/R$, which is closely true. With $L \propto R^2 T_e^4$ this gives $\log L = (20/3) \log T_e$, in quite good agreement with observation (bearing in mind the crudeness of our stellar models).

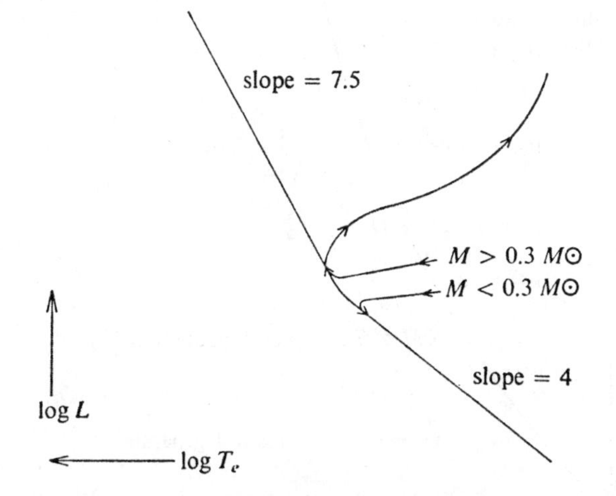

Figure 18. Proto-stars with masses $\leqslant 0.3 M\odot$ contract directly to the white dwarf stage; more massive ones first become main sequence stars, and evolve to become white dwarfs through the red giant stage.

The white dwarfs are distributed over a band of width about 0.5 in log L, having a slope of 4 in agreement with (179). The width of the band is accounted for by the range of masses, from 0.3 to $1.4M\odot$, which are possible in practice. A white dwarf of given mass evolves along a cooling line of slope 4 on figure 17, each mass having its own line.

The upper limit of $1.4M\odot$ for the mass results from (177) with $\mu = 2$, appropriate to a star which has consumed all its hydrogen and helium and synthesised heavier elements from them in earlier phases of evolution.

The lower mass limit results from the high central densities reached in such low mass stars which lead to them becoming degenerate before they have contracted sufficiently for their central temperatures to become high enough for nuclear synthesis – which therefore never becomes available as an energy source. Observation indicates $R \propto M^{0.9}$ on the lower main sequence, equation (6), whereas constant T_c would give $R \propto M$, leading, with M in solar masses, to

$$\rho_c = 100M^{-2} \tag{180}$$

since $\rho_c(\text{sun}) = 100\ \text{g cm}^{-3}$. Equating (180) to (175) gives

$$M = 0{\cdot}17\mu_e^{-5/4}M\odot \tag{181}$$

or, since the unevolved composition has $\mu_e = 0.6$, $M = 0.32M\odot$ as the mass of a star which becomes degenerate at its centre when it

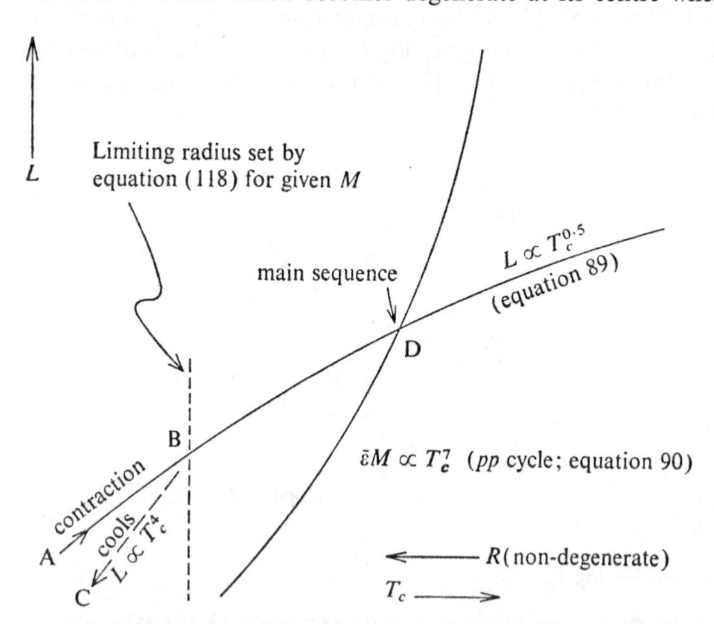

Figure 19. Showing why a low mass star, compelled to follow the track ABC, never becomes a main sequence star represented by point D. Compare to figure 8.

reaches the main sequence after contracting from a low density proto-star. Less massive stars contract straight to the white dwarf stage and therefore the faint end of the main sequence should change slope to 4 from 7.5 at about $\log L = -2$. More massive stars undergo nuclear synthesis and evolve to become red giants, figure 18.

The failure of stars less massive than $0.3M\odot$ to achieve nuclear synthesis is shown on figure 19, which may be compared with figure 8. A star with mass greater than $0.3M\odot$ contracts from A and reaches point D where nuclear energy production by the *pp* cycle balances the luminosity. A star with lesser mass, however, becomes degenerate at some point B before reaching D. At that point the radius has reached the value given by (177), and thereafter it cools at constant radius.

Because of the very high electron densities and momenta, white dwarfs and cores of red giants have very high thermal conductivities and become almost isothermal. Since $T_e \rightarrow T_c$, the cooling line on figure 19 becomes, by (179), $L \propto T_c^4$. The interior temperature never reaches the level required for nuclear synthesis, but the surface temperatures of white dwarfs may become so high that they radiate x-rays.

5.2. *Neutron Stars.*

Equation (178) was reached from the basis that, by the Pauli Exclusion Principle, each electron must have a volume $\frac{1}{2}(\Delta q)^3$ where $\Delta q \,.\, p = \hbar$, p being the range of momentum and the factor two arising because two electrons of opposite spins are not excluded from occupying the same cell of phase space.

Furthermore, the polytropic solution is made on the basis of hydrostatic equilibrium, for which $P(\text{gas}) = P(\text{gravity})$. In the non-relativistic case (77), $P(\text{gas}) \propto \rho^{5/3}$ or, since $\rho \propto M/R^3$,

$$P(\text{gas}) \propto \frac{M^{5/3}}{R^5}$$

Now $P(\text{gravity}) \propto M^2/R^4$ (equation 8) and consequently

$$\frac{P(\text{gas})}{P(\text{gravity})} \propto \frac{M^{-1/3}}{R} \propto M^0$$

using (177). That is, the pressure balance can be achieved at any mass M. On the other hand, in the case of relativistic degeneracy, (81) gives $P(\text{gas}) \propto \rho^{4/3} \propto M^{4/3}/R^4$ whence

$$\frac{P(\text{gas})}{P(\text{gravity})} \propto M^{-2/3}$$

and pressure balance can be achieved at one value of mass only.

Suppose mass is added above the limit set by (178); the self-gravitation now increases so that $P(\text{gravity}) > P(\text{gas})$ and the material is compressed so that the Pauli Exclusion Principle would be violated if the electron density were not prevented from increasing by withdrawal of electrons; this withdrawal of electrons is effected by fusion with protons to form neutrons.

Now from (73), the electron density limit set by the Pauli Exclusion Principle, shown on figure 3, is $n_e = 8\pi p_0^3/3h^3$, and since $p_0^2 = 2mE$ this is

$$n_e = \frac{8\pi}{3} \cdot \frac{(2mE)^{3/2}}{h^3} \tag{182}$$

With $E = 0.8$ MeV, the binding energy of the neutron, this gives $n_e = 3.25 \times 10^{30}$ cm^{-3} (this may be compared with $3/4\pi r_0^3$, where r_0 is the Compton wavelength of the electron; $3/4\pi r_0^3 = 4 \times 10^{30}$ cm^{-3}).

Hence with $\mu_e = 2$ the limiting mass density is

$$\rho = \frac{n_e m_{\mathrm{H}}}{\mu_e} = 2{\cdot}7 \times 10^6 \text{ g cm}^{-3} \tag{183}$$

At densities greater than this, fusion $e^- + p \rightarrow n$ begins, and will continue as the density is increased until each neutron has space $\frac{1}{2}(\Delta q)^3$ where $\Delta q \,.\, p = \hbar$. The limit is reached when Δq is equal to the radius of the neutron; taking this to be the Compton wavelength $r_n = h/m_n c$, the corresponding density is

$$\rho = \frac{m_n}{\frac{4}{3}\pi r_n^3} = 1.7 \times 10^{14} \text{ g cm}^{-3} \tag{184}$$

Until this density is approached, the neutron gas will not be degenerate.

When self-gravitation squeezes the matter to higher densities, the Compton wavelength must be reduced since Δq must be reduced; p cannot be changed because, remembering that pressure is rate of transfer of momentum and that the pressure is determined by the self-gravitation ($P(\text{gravity}) \propto M^2/R^4$), the momentum per neutron is determined by the gravitational self-attraction. The Compton wavelength can only be reduced by increasing the mass of the particle; that is to say, neutrons begin to fuse to form hyperons at densities $> 1.7 \times 10^{14}$ g cm^{-3}.

At first sight the equation of state for a degenerate neutron gas will be similar to (77) and (81) for a degenerate electron gas, with neutrons replacing electrons, so that μ_e becomes m_n/m_{H}: but suppose that $M = 5.75 M\odot$, the limit set by (178); then with $\rho = 1.7 \times 10^{14}$ g cm^{-3},

$$R = \left(\frac{3M}{4\pi\rho}\right)^{1/3} = 25 \text{ km} \qquad (185)$$

and the gravitational potential energy per unit mass is

$$\frac{\Omega}{M} \approx \frac{GM^2}{R} = 3 \times 10^{20} \text{ erg g}^{-1} = \tfrac{1}{3}c^2$$

That is, one third of the rest mass energy is required to dissipate the star! Matter in the star is in an accelerated frame of reference in the sense of general relativity; consequently a general relativistic formulation of the pressure must be used in the equation of state. This modifies the results to give limits to the mass of a neutron star

$$0.05 \leqslant M \leqslant 2.0 M_\odot \qquad (186)$$

analogous to (178) in the case of electron degeneracy. Correspondingly the limits on the radii are $7 \leqslant R \leqslant 9$ km. Stars with masses less than $0.15 M_\odot$ are not, however, expected to become neutron stars because they are bound less tightly than white dwarfs; thus to turn a white dwarf of mass $M < 0.15 M_\odot$ into a neutron star would require work to be done on it by its surroundings. These low mass stars most probably remain as cooling white dwarfs.

Collapse of a star of mass 1.0 $M_\odot$ to a radius of 8 km releases gravitational energy per unit mass of 3×10^{20} erg g^{-1}. If all this energy goes into heating the gas it will raise the temperature to 4×10^{12} K. Because of the high particle momenta, neutron stars will have high thermal conductivities and consequently will be almost isothermal. They will therefore have surface temperatures $\sim 10^{12}$ K; as this state is approached the intense radiation pressure will blow off any surrounding envelope, and the neutron star will cool very rapidly. It may be noted that the total energy content, 10^{54} erg, is comparable with the energy in supernovae explosions. Furthermore, as a result of the extremely high surface temperature the cooling time, which is of the order of the energy content divided by the rate of radiation $4\pi R^2 T_e^4$, is only about 10^3 secs from 10^{12} K to 10^{10} K, and the temperature is down to about 10^8 K in two hundred years. It is in such a short time scale that the neutron star develops the kind of structure described below.

At a temperature of 4×10^{12} K, the neutron velocity

$$v = (3kT/m)^{\frac{1}{2}} = 3 \times 10^{10} \text{ cm s}^{-1} = c.$$

However, not all the gravitational energy released will go into heating the gas: some will be lost by radiation and some by ejection of excess mass as the star evolves to the neutron star state. Consequently the temperature will not reach such a high value and $v < c$, so that the neutron star will not be relativistically degenerate even at the limiting density when fusion to hyperons begins.

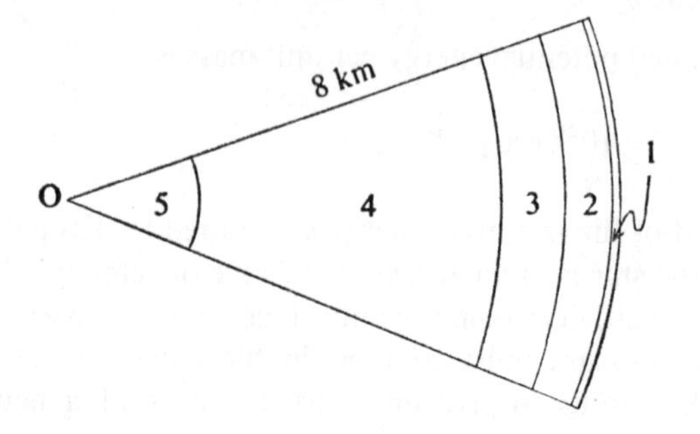

Figure 20. The zones inside a neutron star; their remarkable properties are described in the text.

The structure of a neutron star can now be understood with reference to figure 20; the five zones numbered 1 to 5 have very different properties, as follows:

1. Density 10^4 to 10^5 g cm^{-3}. The enormous magnetic field at the surface causes the atoms to be much more tightly bound and smaller than in a perfect gas. The density is 10^4–10^5 g cm^{-3} and the atoms combine to form a liquid or solid surface. The high electron mobility gives the surface a high thermal and electrical conductivity but the motion of the electrons is constrained along the magnetic lines of force and so the surface takes on the properties of a one-dimensional metal. This layer is only 10^3 cm thick.
2. Density 10^5 to 4×10^{11} g cm^{-3}. The atomic nuclei arrange themselves into a body-centred cubic lattice; all of this region has cooled below the melting temperature in the first two hundred years. The solid material thus formed has a stiffness some 10^{17} times that of steel and an electrical conductivity 10^6 times that of copper!
3. Density 4×10^{11} to 3×10^{14} g cm^{-3}. The lattice arrangement of the nuclei continues downwards through this region, and the lattice Debye temperature is so much higher than the actual temperature that the material is effectively frozen, at 10^{10} K. On the other hand the neutrons form a superfluid, passing freely between the lattice nuclei.
4. Density 3×10^{14} to 8×10^{14} g cm^{-3}. So great has the density become that nearly all the protons and electrons have combined to neutrons (see equation 184) and some of the neutrons have combined to form hyperons; all this region is a superfluid.

5. Density greater than 8×10^{14} g cm^{-3}. The structure is uncertain but as the density rises above 10^{15} g cm^{-3} the neutrons and hyperons may be forced to arrange themselves into a crystal lattice. At still higher densities the core is composed mostly of hyperons.

5.3. *Pulsars.*

The maximum speed of rotation of a neutron star is given by $v(\text{rot}) = (GM/R)^{\frac{1}{2}} = 1.3 \times 10^{10}$ cm s^{-1}, from which it follows that the period of rotation is $T \geqslant 0.4$ ms. In practice speeds of rotation will be less than this as a result of the previous history of the star. During the red giant stage, the great increase in radius will have caused a slowing down of the speed of rotation of the envelope, and although the core contracts, the coupling between core and envelope is sufficiently strong to ensure that most of the angular momentum of the core is transferred to the envelope. The explosion which blows off the envelope when the degenerate core reaches the mass limit given by (178) results in most of the angular momentum being carried away, so that the period of rotation will be substantially larger than that given above.

Pulsars send out pulses of radio emission at intervals ranging from about 30 ms to 4 s. The most generally accepted explanation of the pulses is that they are beams of radio waves emitted by rotating neutron stars after the manner of a lighthouse; the duration of the pulse is about one-twentieth of the period. The explanation requires the presence of very large magnetic fields associated with the neutron star, of the order 10^{12} G. It is outside the scope of this book to consider the mechanisms which have been suggested for the generation of the radio emission; but it is noteworthy that despite the many uncertainties in the theories of pulsars and their behaviour, their discovery has confirmed the existence of the neutron stars predicted by astrophysics.

Explosive Events

PART TWO

and Nuclear Synthesis

CHAPTER SIX

6. Supernovae. 6.1. 6.2. *Introduction to Synthesis of the Elements.* **6.3.** *The Approach to Instability.* **6.4.** *Instability and Explosion.* **6.5.** *Nuclear Synthesis in Stellar Explosions.*

6.1.

In our studies of the structure and evolution of stars in the first part of this book, the condition of hydrostatic equilibrium was the basis of all the models considered; we dealt only with stable stars. There are, however, two ways in which those stellar models may, sooner or later, become explosively unstable.

1. As the mass of the degenerate core grows to exceed the limit for electron degenerate material (equation 178), it must collapse to form a neutron star and ultimately perhaps a baryon star. This collapse releases a large amount of gravitational energy which is probably transferred to the envelope, expanding it to infinity and thus removing the excess mass. Since the mass limit for neutron and baryon stars is less than that for electron degenerate stars, the electron degenerate core may break up into several neutron stars during the collapse and explosion.
2. In degenerate material, the pressure depends on density alone, not on temperature. Therefore if an energy-producing nuclear reaction begins in degenerate material, heating it, there is no rise in pressure and the material does not expand and cool; since the rate of nuclear reactions increases very rapidly with rising temperature, the material heats up until its degeneracy is removed (see description of stage 14 to 15(b) in section 3.2). Then the pressure increases rapidly on account of the high temperature, since now $P \propto T$, and the gas expands explosively.

These two cases may represent the supernovae explosions of type II (population I) and type I (population II) respectively. Nuclear fusion before and during these explosions results in the building up of heavier elements from hydrogen and their ejection back into the interstellar gases from which new stars form. Therefore later generations of stars are expected to have higher abundances of these heavier elements. There is evidence in the strengths of the lines due to these elements in the spectra of stars of various ages that abundances of heavy elements have increased with time during the life of the Galaxy by a factor of a thousand or more.

6.2. *Introduction to Synthesis of the Elements.*

The synthesis of helium, carbon, nitrogen and oxygen have already been dealt with in considering the sources of energy of a star during evolution on the main sequence and in the red giant stage:

T(K)	*Reaction*	*Products*
$\leqslant 1.8\times 11^7$	pp	${}_2He^4$
$\geqslant 1.8\times 10^7$	CN	${}_2He^4$
$>10^8$	3α	${}_6C^{12}$

C^{13}—N^{14}—N^{15}

${}_7N^{14}(\alpha,\gamma){}_9F^{18}(\beta^+,\nu_+){}_8O^{18}$

The ${}_6C^{12}$ required in the CN cycle must have been produced by the 3α cycle in earlier generations of stars.

As a star exhausts each nuclear fuel in turn, it contracts and heats up (the core does, even if the envelope expands), drawing on its supply of gravitational energy until the interior temperature has risen high enough for the next nuclear reaction to take place. After the synthesis of carbon by the 3α process, the following reactions occur in turn as the temperature rises:

a) $2\times 10^8 \leqslant T \leqslant 6\times 10^8$ K. In this range of temperatures four kinds of process occur:

(i) ${}_{10}Ne^{22}(\alpha\gamma, n){}_{12}Mg^{25}$ (endothermic α-process)

(ii) $C^{12}(\alpha,\gamma)O^{16}$ (exothermic α-process)

(iii) $C^{12}(C^{12},\gamma)Mg^{24}$

The first of these produces neutrons which are then available for:

(iv) slow neutron capture processes by elements E (the s-process)

$${}_ZE^A(n,\gamma){}_ZE^{A+1}$$
$${}_ZE^{A+1}(\beta^-\nu_-){}_{Z+1}E^{A+1}(n,\gamma){}_{Z+1}E^{A+2}\ldots$$

This is called the slow or s-process because neutron capture occurs at a rate slow in comparison with β-decay. It may be responsible for producing the majority of nuclei in the range $23 \leqslant A \leqslant 46$.

Now in equilibrium, rate of synthesis $E^{A-1}\rightarrow E^A$ = rate $E^A\rightarrow E^{A+1}$ = etc. That is, if N denotes abundance and σ denotes neutron capture cross-section,

$$N(E^{A-1})\sigma_{A-1} = N(E^A)\sigma_A = N(E^{A+1})\sigma_{A+1} = \text{etc.},$$

or

$$N(E^A) \propto \frac{1}{\sigma_A}$$

Thus in the case of the s-process, equilibrium abundances are inversely proportional to neutron capture cross sections.

Note that $\sigma \propto (1/m_A v_A^2 + 1/m_n v_n^2)$ and so N will vary with A for equipartition of energy and will vary similarly for all A if v_n is varied.
b) $T = 2 \times 10^9$ K. The higher energies of collision result in synthesis of sulphur from oxygen:

$$O^{16}(O^{16}, \gamma)S^{32}$$

c) $T \geqslant 3 \times 10^9$ K. The reactions occurring at this temperature are considered in the following section.

6.3. *The Approach to Instability.*

The temperature is now so high that many different reactions are taking place at once, viz:

synthesis (α, γ), (p, γ), (n, γ), (p, n)
dissociation (γ, α), (γ, p), (γ, n)

These are occurring so quickly that the abundances are in statistical equilibrium. Suppose a nucleus (A, Z) is in equilibrium with the ambient neutrons and protons; then

$$n(A, Z) = g(A, Z)\left(\frac{2\pi A m_H kT}{h^2}\right)^{3/2} \frac{n_n^{A-Z} n_p^z}{Z^A} \times \left(\frac{h^2}{2\pi mkT}\right)^{3A/2} \exp\frac{Q(A, Z)}{kT} \tag{187}$$

where $g(A, Z) = \sum_r (2I_r + 1)\exp(-E_r/kT)$ is the statistical weight, I_r being the spins and E_r the energy above the ground level of an excited state, and $Q(A, Z) = c^2[(A-Z)m_n + Zm_p - m(A, Z)]$ is the binding energy. Equation (187) can be derived simply as follows.

If p is the range in relative momentum of a neutron and a nucleus, then the number of available momentum cells per unit volume permitted by the Pauli exclusion principle is $n_m = 1/(\Delta q)^3$ where $p \, . \, \Delta q = \hbar$. Hence

$$n_m = \frac{p^3}{\hbar^3} = \frac{(2\pi p)^3}{h^3} = \frac{[2\pi(kT/2\pi m)^{\frac{1}{2}}]^3}{h^3}$$

for a Maxwellian distribution of velocities; thus

$$n_m = \frac{(2\pi mkT)^{3/2}}{h^3} \tag{188}$$

If g quantum states are permitted in each momentum cell, then the number of available quantum states per unit volume is

$$n_q = \frac{g(2\pi mkT)^{3/2}}{h^3} \tag{189}$$

Let the number of particles of a given kind per unit volume be denoted by n; then the number of quantum states available to each particle of this kind is

$$\nu = \frac{n_q}{n} = \frac{g(2\pi mkT)^{3/2}}{nh^3} \tag{190}$$

The probability w of the particle being in any one of them, assuming that they are all equally probable, is

$$w = \frac{1}{\nu} = \frac{nh^3}{g(2\pi mkT)^{3/2}} \tag{191}$$

Also, for a large number of particles, the fraction in a given state is equal to the probability of a single particle being in that state. Now the Boltzmann Law gives

$$S = \frac{Q}{T} \propto \log\left(\frac{w}{w_0}\right) = -k\log\left(\frac{w}{w_0}\right) \tag{192}$$

where k is the Boltzmann constant of proportionality, Q is the energy of the state at temperature T and w_0 is the probability of occupying the zero energy state (ground level). The negative sign arises because the probability of zero temperature or infinite energy must be zero and the logarithm of zero is negative infinity. Thus,

$$\log\left(\frac{w_1}{w_2}\right) = -\frac{(Q_1 - Q_2)}{kT} \tag{193}$$

This will be used later.

If there are $n(A, Z)$ particles of type (A, Z) per unit volume, the probability of one of them being in a particular quantum state is, from (191),

$$w(A,Z) = \frac{n(A,Z)h^3}{g(A,Z)(2\pi AmkT)^{3/2}} \tag{194}$$

Let the number of neutrons per unit volume be denoted by n_n; the probability of one occurring in the same volume of phase space as the particle (A, Z), (that is, moving in the same direction with the same momentum at the same position, within the limits set by $p\,.\,\Delta q = \hbar$), and having the same spin, is, by (191),

$$w(n) = \frac{n_n h^3}{g(n)(2\pi mkT)^{3/2}} \tag{195}$$

The probability of two coincidences in the given state is $(w(n))^2$; of three, $(w(n))^3$; of $(A-Z)$ coincidences, $(w(n))^{A-Z}$. Similarly for protons, the probability of Z coincidences is $(w(p))^Z$.

If follows that the probability of $(A-Z)$ neutrons and Z protons being in the given state corresponding to that of (A, Z) is

$$w(n)^{A-Z}w(p)^Z = \frac{n_n^{A-Z}n_p^Z}{g_n^{A-Z}g_p^Z}\left(\frac{h^2}{2\pi mkT}\right)^{3A/2} \tag{196}$$

But, by (192),

$$\frac{w(A,Z)}{w(n)^{A-Z}w(p)^A} = e^{-Q/kT} \tag{197}$$

where $Q = Q_{AZ} - (n_nQ_n + n_pQ_p) = c^2[Am - ((A-Z)m_n + Zm_p)]$. Noting that since $g_n = g_p = 2$, $g_n^{A-Z}g_p^Z = 2^A$, it follows from (194) and (197) that

$$n(A,Z) = g(A,Z)\left(\frac{2\pi AmkT}{h^2}\right)^{3/2}\left(\frac{n_n^{A-Z}n_p^Z}{2^A}\right) \times$$

$$\times (2\pi mkT)^{-3A/2}e^{-Q/kT} \tag{198}$$

which is identical with (187). This gives the abundances of elements (A, Z) when the synthesis and dissociation processes are occurring so quickly that the abundances are in statistical equilibrium.

If the time scale within which these processes are operating is long in comparison with β-decay times, then the relative total numbers of neutrons and protons are determined by β-decay. If the time scale is comparatively short, then they are equal to the initial values which are determined by the initial composition. Since the energies Q and statistical weights g are known, the numbers $n(A, Z)$ of the various elements (A, Z) can then be calculated for a range of densities and temperatures (density $= \sum n(A, Z)Am$ since the mass contribution by free neutrons and protons is small).

These are the processes which result in the building up of the iron group of elements from Mg^{24}, equilibrium abundances being set up over the range of elements from titanium to copper.

6.4. *Instability and Explosion.*

A particular interesting situation now develops as the temperature rises above 5×10^9 K. The equilibrium between an atom (A, Z) and its products of dissociation a α-particles and b neutrons is given, similarly to (198), by

$$n(A,Z) = \frac{g(A,Z)}{g_\alpha^a g_n^b}\left(\frac{A}{A_\alpha^a A_n^b}\right)^{3/2} n_\alpha^a n_n^b \left(\frac{h^2}{2\pi mkT}\right)^{3(a+b-1)/2} \times e^{-Q/kT} \tag{199}$$

For $Fe^{56} \rightleftharpoons 13\alpha + 4n$, $g(A, Z) \simeq 1.4$, $Q = 124.4$ MeV, $g_\alpha = 1$, $g_n = 2$, $A_\alpha = 4$, $A_n = 1$, $a = 13$, $b = 4$, $A = 56$. It follows that

$$n(\mathrm{Fe}^{56}) = 1.4\,\frac{56^{3/2}}{2^{43}}\,n_\alpha^{13}n_n^4\left(\frac{h^2}{2\pi mkT}\right)^{24} e^{-14.5/T_9} \tag{200}$$

where T_9 is the temperature in units of 10^9 K. Now using this equation to calculate the relative amounts of mass in Fe^{56}, and in α-particles and neutrons as the temperature and density are varied (noting that $n_\alpha/n_n = 13/4$ in equilibrium with iron), it turns out that half the mass is in Fe^{56} and half in α-particles and neutrons when the density is given by

$$\log \rho = 11.62 + 1.5 \log T_9 - 39.17/T_9 \tag{201}$$

ρ being in g cm^{-3}. This relationship is shown on figure 21. It marks the transition between material consisting mostly of iron and material composed mostly of α-particles and neutrons. The change from Fe^{56} to $13\alpha + 4n$ occurs extremely rapidly with increasing temperature, as can be seen by reference to (200) where $n(\mathrm{Fe}^{56})$ varies as

$$n(\mathrm{Fe}^{56}) \propto T_9^{-24}\,e^{-14.5/T_9}$$

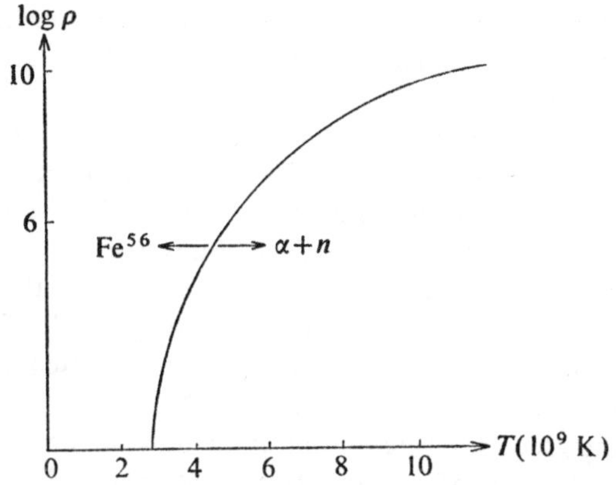

Figure 21. Implosion imminent. On one side of the boundary the gas forms mostly into iron nuclei, on the other side it breaks up into α-particles and neutrons. The difference in the binding energies is greater than the total heat content of the whole star.

Now the binding energy of the thirteen α-particles and four neutrons in Fe^{56} is 124.4 MeV, that is, 2×10^{18} erg g^{-1}. The dissociation $\mathrm{Fe}^{56} \rightarrow 13\alpha + 4n$ withdraws this much heat from the gas. But the heat content of the gas is $U = C_V T = 3\mathscr{R}T/2\mu$ g^{-1}. In this, the molecular weight is 1.6, since there are 112 mass units made up of thirteen α-particles, four neutrons and one iron nucleus, and 70 particles made up of 26 'iron' electrons, 26 'α-particle' electrons, thirteen α-particles, four neutrons and one iron nucleus. Thus even at $T_{,9} = 10$the heat content is only 8×10^{17} erg g^{-1}.

Consequently as the temperature exceeds the critical value given by (201) the dissociation of iron withdraws all the thermal energy from the gas, which therefore cools, and collapses through the resulting disappearance of the supporting gas pressure. This situation develops in the core of the star, but the collapse of the core removes the pressure support of the envelope which also collapses inwards, rapidly heats up, and explodes due to ignition of nuclear fuel.

The collapse of the core releases a large amount of gravitational energy. The explosion of the envelope produces the spectacular event which we know as a supernova; the collapsed core remains behind as a star, later to be observed perhaps as a pulsar in the expanding gaseous remnant of the supernova explosion.

6.5. *Nuclear Synthesis in Stellar Explosions.*

Only nuclear reactions that occur very rapidly are of significance in the short time scales involved in a stellar explosion, which is of the order of 100 seconds. β-decay times are much longer than this but neutron capture can occur more quickly. These rapid neutron captures, the so-called *r*-process, are of the form $(n, \gamma) \rightleftharpoons (\gamma, n)$ and are in equilibrium for a short time; as the elements build up by neutron capture there is a slow leakage by β-decay.

The relative abundances of the elements produced are calculated as follows. From (194), we can write for the equilibrium situation

$$\frac{w(A+1,Z)}{w(A,Z)w(n)} = \frac{n(A+1,Z)h^3}{g(A+1,Z)(2\pi(A+1)mkT)^{3/2}} \times$$
$$\times \frac{g(A,Z)g(n)(2\pi mkT)^3 A^{3/2}}{n(A,Z)n(n)h^6}\left(\frac{m(n)}{m}\right)^{3/2} \quad (202)$$

But from (193),

$$\frac{w(A+1,Z)}{w(A,Z)w(n)} = e^{-Q/kT} \quad (203)$$

where

$$Q = Q(A+1,Z) - (Q(A,Z)+Q(n)) \quad (204)$$

Combining (202) and (203) gives

$$\frac{n(A+1,Z)}{n(A,Z)} =$$
$$n(n)\frac{h^3}{(2\pi mkT)^{3/2}} \cdot \frac{g(A+1,Z)}{g(A,Z)g(n)}\left(\frac{A+1}{A}\right)^{3/2}\left(\frac{m}{m(n)}\right)^{3/2} \times$$
$$\times e^{-[Q(A+1,Z)-(Q(A,Z)+Q(n))]/kT} \quad (205)$$

This shows that the ratio of the abundances of successive elements,

$$\log\left(\frac{n(A+1,Z)}{n(A,Z)}\right) \tag{206}$$

is a maximum when

$$Q(A,Z)-Q(A+1,Z) \tag{207}$$

is a maximum. The function (207) is plotted against neutron number $(A-Z)$ in figure 22, showing that sharp maxima occur when there are closed neutron shells, giving corresponding abundance maxima at values of $(A-Z)$ of 50, 82 and 126. These are unstable nuclei, and decay by β-emission at constant atomic weight until they reach stable configurations. There is some evidence that the material ejected by supernovae explosions has been affected by these processes.

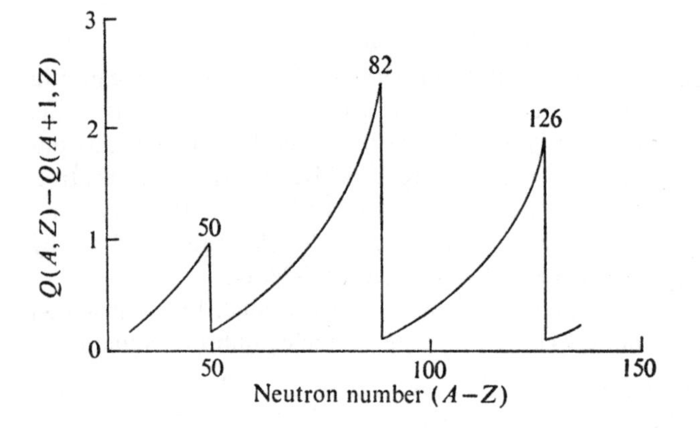

Figure 22. Abundances of the elements, peaking at closed neutron shells.

7.1. *Mass Limits for Stable Main Sequence Stars.*

The first part of this book dealt with the structure and evolution of stable stars, but it did not consider if there was a limit to the mass which could form a stable star. The condition for stability was found to be

$$3(\gamma-1)U+\Omega=0 \qquad (50)$$

Let us assume that throughout the star the ratio of radiation pressure to gas pressure is constant; thus

$$P_g = \beta P = \frac{\rho \mathcal{R} T}{\mu}; \qquad P = \frac{\rho \mathcal{R} T}{\mu\beta} \quad \text{and} \qquad (208)$$

$$P_r = (1-\beta)P = \frac{aT^4}{3}; \qquad P = \frac{aT^4}{3(1-\beta)} \qquad (209)$$

Note that this is a polytrope with index 3, because since $P \propto \rho T$ and $P \propto T^4$ it follows that $\rho \propto T^n$ where $n=3$ (see p. 16).

The internal energy is

$$U = \tfrac{3}{2}PV = \frac{3\mathcal{R}T}{2\beta}\ \text{mol}^{-1} \qquad (210)$$

and the gravitational energy is

$$\Omega = -q\frac{\mu GM}{R}\ \text{mol}^{-1} \qquad (211)$$

Since for a polytrope of index 3 (see p. 16) $q \sim 3/2$, the condition for stability becomes

$$\beta > \frac{3(\gamma-1)\mathcal{R}TR}{\mu GM} \qquad (212)$$

For example, consider main sequence stars, which are well represented by a polytrope of index 3. For them we have $M/R \approx$ constant $= M\odot/R\odot = 3\times10^{22}$; $\bar{T} = 0.4T_c \approx 8\times10^6$ K; $\mu = 0.6$; $4/3 \leqslant \gamma \leqslant 5/3$. Putting these values into (212) gives $\beta > 0.6$ for stability. That is, gas pressure must be more than 60 per cent of the total pressure, and radiation pressure correspondingly must be less than 40 per cent of the total pressure; if it is more than this, radiation pressure will blow the mass away. Now it has already been shown (equation 60) that radiation pressure increases relative to gas pressure with increasing stellar mass; this suggests that there may be an upper limit to the mass of a star, above which gravitational attraction is not powerful enough to retain material against the outward pressure of the star's own radiation. Now eliminating T from (208) and (209) gives

$$P = K\rho^{4/3} \tag{213}$$

where

$$K = \left(\frac{3\mathscr{R}^4(1-\beta)}{a\mu^4\beta^4}\right)^{1/3} \tag{214}$$

Equation (213) represents a polytrope of index $n=3$ for which, from (173) and (174),

$$M = 4\pi\left(\frac{K}{\pi G}\right)^{3/2}\left(-z^2\frac{du}{dz}\right)_{u=0} \tag{215}$$

Tables of the solutions of the equations of polytropes give, for $n=3$, $(-z^2\,du/dz)_{u=0} = 2.0$, and hence (215) gives

$$K = \frac{\pi^{1/3}G}{4}M^{2/3} \tag{216}$$

Equating (214) to (216) gives the mass in terms of $\beta = P_g/P$ as

$$M = \frac{8(3/\pi a G^3)^{\frac{1}{2}}(\mathscr{R}^2/\mu^2)(1-\beta)^{\frac{1}{2}}}{\beta^2}$$

or

$$\frac{M}{M\odot} = \frac{18(1-\beta)^{\frac{1}{2}}}{\mu^2\beta^2} \tag{217}$$

But for stability β must be greater than the right-hand side of (212). For main sequence stars $\beta > 0.6$ as shown above, and (216) then gives

$$M < 114M\odot \tag{218}$$

This is the most massive main sequence star that can be formed. If more mass is added to the star, the radiation pressure increases and blows it off. This can be seen by reference to the equation of hydrostatic equilibrium

$$\frac{dP}{dr} = -\rho\frac{GM}{r^2} \tag{31}$$

When the mass is very large, the pressure is mainly radiation pressure (60); but along the main sequence $M/R \sim$ constant and so $\rho M/r^2 \propto M/R^4 \propto 1/M^3$ at any given r/R. That is to say, the gravitational self-attraction decreases sharply, as $1/M^3$, up the main sequence, and when $M > 114 M_\odot$ it has become so low that it can no longer hold the mass together against the gradient of radiation pressure.

7.2. *Supermassive Objects.*

Suppose, however, that the gravitational potential energy is comparable with the rest mass energy, that is,

$$\Omega = \frac{3}{2}\frac{GM^2}{R} \sim Mc^2 \qquad (219)$$

or, writing $R = R_g$ when this condition obtains,

$$R_g = \frac{3}{2}\frac{GM}{c^2} = 10^{-28} M \qquad (220)$$

The mean density of such stars is

$$\bar{\rho} = \frac{M}{\frac{4}{3}\pi R_g^3} = 4.5 \times 10^{16} \left(\frac{M_\odot}{M}\right)^2 \qquad (221)$$

For example, $M = 10^8 M_\odot$, $R_g = 314 R_\odot$, $\bar{\rho} = 4.5$ g cm^{-3}. Gravitational self-attraction $\sim -\rho GM/r^2 \propto M^2/R^5 = 3000$ solar units.

Compare this with a massive main sequence star for which $M = 100 M_\odot$, $R = 100 R_\odot$, $\bar{\rho} = 10^{-4}$ g cm^{-3}, $M^2/R^5 = 10^{-10}$. Thus when the gravitational potential energy becomes comparable with the rest mass energy Mc^2, the gravitational self-attraction becomes much stronger again and binds the mass tightly together.

7.3. *Hydrostatic Equilibrium.*

In such supermassive objects we must take account of additional factors in the equation of hydrostatic equilibrium: we must (a) include the mass equivalent of the energy (especially the radiation energy density) in both the density and the mass; and (b) use for the radial distance the general relativistic coordinate. Two things are combining to make it more difficult for radiation pressure to blow off mass: (i) the mass equivalent of the energy is considerable and increases the gravitational attraction on an atom; and (ii) the photons lose energy in doing work against the strong gravitational field and have less energy remaining to do work on atoms.

To add the mass equivalent of the energy to the density in the equation of hydrostatic equilibrium, we add to $\rho\, dr$ the mass equivalent of the work done by the pressure P in traversing the distance

dr, which is Pdr/c^2, so that

$$\rho dr \text{ becomes } \rho dr + \frac{P dr}{c^2} \tag{222}$$

The volume element $V = 1 \times 1 \times dr$ is to be replaced by the General Relativistic volume element:

$$V \text{ becomes } V \Big/ \left(1 - \frac{2GM}{rc^2}\right)^{\frac{1}{2}}$$

and hence

$$dr \text{ becomes } \frac{dr}{(1 - 2GM/rc^2)^{\frac{1}{2}}} \tag{223}$$

The total mass interior to r has now to have the mass equivalent of the energy added and the volume converted, thus

$$M_r = \int_0^r 4\pi\rho r^2 \, dr \text{ becomes}$$

$$M(\text{total})_r = M_r \left(1 + \frac{4\pi\alpha(r) P r^3}{M_r c^2}\right) \Big/ \left(1 - \frac{2GM_r}{rc^2}\right)^{\frac{1}{2}} \tag{224}$$

where $\alpha(r) = P/\bar{P}$, $\bar{P}$ being the mean pressure interior to r. Thus finally, (222), (223) and (224) give a new general relativistic equation of hydrostatic equilibrium,

$$\frac{dP}{dr} = -\frac{\rho G_{eff} M_r}{r^2} \tag{225}$$

where

$$G_{eff} = G\left(1 + \frac{P}{\rho c^2}\right)\left(1 + \frac{4\pi\alpha(r) P r^3}{M_r c^2}\right) \Big/ \left(1 - \frac{2GM_r}{rc^2}\right) \tag{226}$$

is the effective gravitational constant.

Direction and distance in the region of supermassive objects are changed markedly due to the effect of the gravitational field on the light paths, figure 23. If the mass M_r is small, the light path (a) is essentially a straight line; but when M_r becomes so large that condition (220) applies the gravitational attraction on the mass equivalent of the photon energy becomes considerable and the light path becomes curved, (b). The curvature r of the surface of M_r is measured relative to the light path – which is the conventional straight line – and relative to light path (b) the curvature of the surface is much less than it would be measured from (a). Thus, regarded in this conventional way, r *appears* to get larger as M_r approaches the critical value. When $M/r = 2c^2/3G$ the radius of curvature of the light path

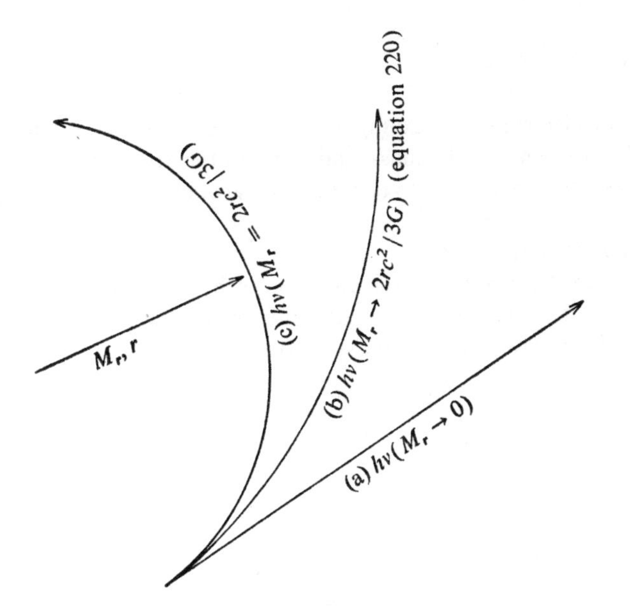

Figure 23. The paths of photons close to objects of low mass (a), and to supermassive objects (b), (c).

is also r; the curvature of the surface relative to the light path is then zero, r *appears* to be infinite and so does the volume. Photons are trapped by the gravitational field, like planets around the Sun. Unable to emit light, the object has created a black hole in space. The effective gravitational constant $G_{\overline{eff}}$, (226), becomes infinite. This explanation of the properties of a supermassive object is a partly Newtonian view and not wholly justifiable but nevertheless gives some working understanding of what is happening.

We began this chapter by finding that radiation pressure on atoms is stronger than the gravitational attraction on them for main sequence stars with masses much above 100 $M\odot$, but then found that if the mass is made very large, say $\sim 10^8$ $M\odot$, the effective gravitational constant is increased enormously by the mass equivalent of the energy, and the effect of radiation pressure is much reduced because the photons lose energy moving against the gravitational field (and at a critical mass have insufficient energy to escape at all). It follows that the object can remain gravitationally bound even if the temperature of the material becomes much higher than it does in a main sequence star; it may, in fact, become so high that the speed of motion of the particles approaches that of light, that is:

$$\tfrac{3}{2}kT \gtrsim \tfrac{1}{2}mc^2; \quad \text{or} \quad T \gtrsim \frac{mc^2}{3k} \tag{227}$$

(In the case of electrons, for example, this situation occurs when $T \geqslant 2 \times 10^9$ K.) In this case the mass equivalent of the kinetic energy of the particles would also have to be taken into account.

7.4. *Gravitational Energy.*
Corresponding to the changes in the equation of hydrostatic equilibrium (225), similar changes have to be made to the calculation of the gravitational energy of a supermassive object. Thus:

$$-\Omega = \int_0^R \frac{GM_r\, dM_r}{r} = \int_0^R \frac{4\pi G\rho M_r r^2\, dr}{r} \tag{228}$$

becomes

$$\Omega(\text{relativistic}) =$$

$$\int_0^R 4\pi G\rho\left(1+\frac{P}{c^2\rho}\right)\frac{M_r}{r}\left(1+\frac{4\pi\alpha(r)Pr^3}{M_r c^2}\right)\frac{r^2\, dr}{\left(1-\dfrac{2GM_r}{rc^2}\right)^{\frac{1}{2}}} \tag{229}$$

Since $2GM_r/rc^2 < 1$ this can be expanded, giving

$$\Omega(\text{rel}) = G\int_0^R \frac{M_r\, dM_r}{r} + \frac{4\pi G}{c^2}\int_0^R M_r Pr\, dr +$$

$$+\int_0^R \frac{4\pi G}{rc^2}\left(\tfrac{4}{3}\pi(3\alpha(r)P)r^3\right)\rho r^2 dr +$$

$$+\frac{4\pi G^2}{c^2}\int_0^R M_r^2\rho\, dr + O(c^4) \tag{230}$$

The third integral

$$\int_0^R \frac{4\pi G}{rc^2}\left(\tfrac{4}{3}\pi(3\alpha(r)P)r^3\right)\rho r^2\, dr$$

$$= \int_0^R \frac{G\left(\dfrac{\frac{4}{3}\pi r^3(3\alpha(r)P)}{c^2}\right)4\pi\rho r^2\, dr}{r}$$

$$= \int_0^R \frac{GM(P)\, dM_r}{r} \tag{231}$$

where $M(P)$ is the mass equivalent of the energy density due to the pressure interior to r. Now

$$\int_0^R \frac{GM(P)\,dM_r}{r} \tag{232}$$

is the mutual gravitational potential energy due to the integral from $r=0$ to $r=R$ of the mutual gravitational attraction of $M(P)$ and dM_r; this must be the same as the corresponding integral of the mutual gravitational attraction of M_r on $dM(P)$, which is

$$\int_0^R \frac{GM_r(4\pi r^2(P/c^2)\,dr)}{r} \tag{233}$$

Equation (230) then becomes

$$\Omega(\mathrm{rel}) = \Omega + \frac{8\pi G}{c^2}\int_0^R M_r P r\,dr + \frac{4\pi G^2}{c^2}\int_0^R M_r^2 \rho\,dr \tag{234}$$

7.5. *The Total Energy.*

The total energy of a main sequence star is

$$E = \int_0^R U\,dV - \int_0^R \frac{GM_r}{r}\rho\,dV = \int_0^R U\,dV - \Omega. \tag{235}$$

Now

$$\begin{aligned}\int U\,dV &= \int 3P_r\,dV + \int \tfrac{3}{2}P_g\,dV \\ &= \int 3(1-\beta)P\,dV + \int \tfrac{3}{2}\beta P\,dV \\ &= \int 3\left(1-\frac{\beta}{2}\right)P\,dV \end{aligned} \tag{236}$$

Therefore

$$E = \int_0^R 3\left(1-\frac{\beta}{2}\right)P\,dV - \Omega \tag{237}$$

For stability, by (47), $2\mathrm{KE}+\mathrm{PE}=0$. The kinetic energy is $\int_V \frac{3}{2}P\,dV$ and the potential energy is $-\Omega$; hence we have

$$\Omega = \int_V 3P\,dV \tag{238}$$

and the total energy of a stable main sequence star becomes

$$E = -\beta\Omega/2 \tag{239}$$

In the case of a supermassive object, in hydrostatic equilibrium, the energy required to disperse it to infinity with zero temperature is

$$E(\text{rel}) = (M - M_0)c^2 \tag{240}$$

where M_0 is the rest mass and M is the total mass including the mass equivalent of the internal energy when dispersed; thus

$$M_0 = \int_0^R \rho_0 \left(1 - \frac{2GM}{rc^2}\right)^{-\frac{1}{2}} dV \tag{241}$$

and

$$M = \int_0^R \left(\rho_0 + \frac{U}{c^2}\right)\left(1 - \frac{2GM}{(r\to\infty)c^2}\right)^{-\frac{1}{2}} dV$$

$$= \int_0^R \left(\rho_0 + \frac{U}{c^2}\right) dV \tag{242}$$

Writing

$$\rho = \rho_0 + \frac{U}{c^2} \tag{243}$$

for the total density ρ, (240) to (243) give

$$E(\text{rel}) = \int_0^R \rho c^2\, dV - \int_0^R \left(\rho - \frac{U}{c^2}\right)\left(1 - \frac{2GM}{rc^2}\right)^{-\frac{1}{2}} c^2\, dV$$

$$= \int_0^R U\left(1 - \frac{2GM}{rc^2}\right)^{-\frac{1}{2}} dV +$$

$$+ \int_0^R \rho c^2 \left[1 - \left(1 - \frac{2GM}{rc^2}\right)^{-\frac{1}{2}}\right] dV$$

$$= \int_0^R U\left(1 + \frac{GM}{rc^2}\right) dV +$$

$$+ \int_0^R \rho c^2 \left[1 - \left(1 + \frac{GM}{rc^2} + \frac{3}{2}\frac{G^2M^2}{r^2c^4} + \cdots\right)\right] dV$$

$$= \int_0^R U\left(1 + \frac{GM}{rc^2}\right) dV - \int_0^R \frac{\rho GM}{r}\, dV -$$

$$- \int_0^R \frac{3}{2}\frac{\rho G^2M^2}{r^2c^2}\, dV + \text{terms in} \left(\frac{1}{c^4}\right)$$

$$= \int_0^R U\left(1 + \frac{GM}{rc^2}\right) dV - \int_0^R \frac{GM\, dM}{r} - \frac{3}{2}\frac{G^2}{c^2}\int_0^R 4\pi\rho M^2\, dr$$

$$= \int_0^R U\left(1+\frac{GM}{rc^2}\right)dV - \Omega - \frac{6\pi G^2}{c^2}\int_0^R \rho M^2\,dr \qquad (244)$$

When $GM/r \ll c^2$ the last term $\to 0$ and (244) reduces to (235), the total energy of an ordinary star.

The first term on the right-hand side of (244), which for convenience we can denote $U(\text{rel})$, is not yet in a form in which it can be calculated but it can be expressed in terms of M and R. Using (236),

$$U(\text{rel}) = \int_0^R U\left(1+\frac{GM}{rc^2}\right)dV$$

$$= \int_0^R 3\left(1-\frac{\beta}{2}\right)\left(1-\frac{GM}{rc^2}\right)P\,dV \qquad (245)$$

Now

$$\int_0^R P\,dV = \int_0^R d(PV) - \int_0^R V\,dP = [PV]_0^R - \int_0^R V\,dP$$

But $P=0$ at $r=R$, $V=0$ at $r=0$; hence $[PV]_0^R = 0$. Consequently

$$\int_0^R P\,dV = -\int_0^R V\,dP \qquad (246)$$

Substituting (246) in (245) gives

$$U(\text{rel}) = -\int_0^R 3\left(1-\frac{\beta}{2}\right)\left(1-\frac{GM}{rc^2}\right)V\frac{dP}{dr}\,dr \qquad (247)$$

where dP/dr is given by equation (225). After some reduction, following similar lines to that for Ω in (229) to (234), there results

$$\int_0^R U\left(1+\frac{GM}{rc^2}\right)dV =$$

$$\left(1-\frac{\beta}{2}\right)\left[\Omega + \frac{8\pi G}{c^2}\int_0^R MPr\,dr + \frac{12\pi G^2}{c^2}\int_0^R \rho M^2\,dr\right] \qquad (248)$$

Substituting (248) into (244) gives for the total energy

$$E(\text{rel}) = -\frac{\beta\Omega}{2} + \frac{8\pi G}{c^2}\int_0^R M_r Pr\,dr + \frac{6\pi G^2}{c^2}\int_0^R \rho M_r^2\,dr \qquad (249)$$

Remembering that $P \propto M^2/R^4$ and $\rho \propto M/R^3$, each of the integral terms is proportional to G^2M^3/c^2R^2. For the polytrope $n=3$ (equation 213) the integrals can be evaluated using the tabulated functions; and since for $n=3$, $\Omega = 3GM^2/2R$, the total energy becomes

$$E(\text{rel}) = -\frac{3\beta}{4}\frac{GM^2}{R} + 5.1\frac{G^2M^3}{R^2c^2}$$

or more informatively

$$\frac{E(\text{rel})}{Mc^2} = -\frac{3\beta}{4}\left(\frac{GM}{Rc^2}\right) + 5.1\left(\frac{GM}{Rc^2}\right)^2 \tag{250}$$

Equation (250) is the total energy of a supermassive object *in hydrostatic equilibrium.* When $GM/R \ll c^2$ this is negative but as GM/R approaches c^2 the total energy becomes more and more positive and consequently energy must be *added* to the system to maintain equilibrium!

It is instructive to examine the possible evolution of such an object in terms of its central temperature. For a polytrope of index 3,

$$T_c = \left((n+1)z\frac{du}{dz}\right)^{-1}\frac{G}{\mathscr{R}}\mu\beta\frac{M}{R} = 0.86\frac{G}{\mathscr{R}}\mu\beta\frac{M}{R} \tag{251}$$

From (217), when $\beta \ll 1$

$$\mu\beta = \left(\frac{18M\odot}{M}\right)^{\frac{1}{2}} \tag{252}$$

Equations (251) and (252) give

$$R = \frac{5.8 \times 10^{18}}{T_c}\left(\frac{M}{M\odot}\right)^{\frac{1}{2}} \tag{253}$$

Substituting for β and R in (250) using (252) and (253) gives finally

$$\frac{E(\text{rel})}{Mc^2} = -\frac{8.2}{\mu}10^{-14}T_c + 3.3 \times 10^{-27}T_c^2\frac{M}{M\odot} \tag{254}$$

The total energy is shown on figure 24 as a function of T_c for various masses.

It has already been seen, in equations (61) to (66), that as a star contracts releasing gravitational energy, a fraction $(3\gamma-4)/3(\gamma-1)$ of it must be radiated away and a fraction $1/3(\gamma-1)$ retained to increase the internal energy if equilibrium is to be maintained. That is the non-relativistic situation. Now (254) and figure 24 show a contrary state of affairs in the relativistic case; as a supermassive object contracts and its central temperature rises, a point is reached at which large amounts of energy must be *added* if hydrostatic equilibrium is to be maintained, the energy released from gravitation ($\sim Mc^2$, see p. 97) being insufficient to heat up the gas enough for the gas pressure to balance the relativistic gravitational attraction. The energy released by nuclear reactions is much less than the

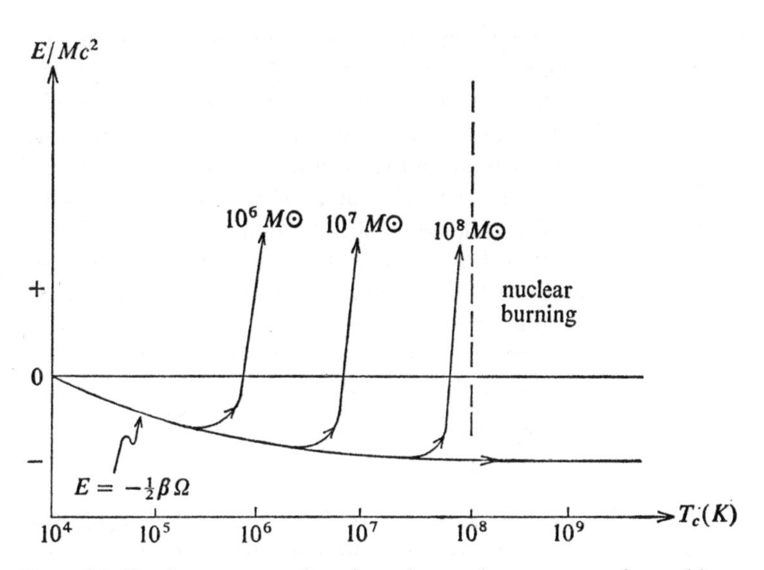

Figure 24. Total energy as a function of central temperature for stable supermassive objects. Since the energy becomes positive as the temperature rises energy must be *added* to the system if it is to remain stable!

energy equivalent of the total mass and is therefore totally inadequate. However, kinetic energy of mass motions in an ensuing collapse may be enough if the velocities approach that of light; this possibility can be taken into account by writing for (225)

$$\frac{dP}{dr} = -\rho\left(\frac{G_{eff}M_r}{r^2}+\frac{dv}{dt}\right) = -\rho(1+f)\frac{G_{eff}M_r}{r^2} \tag{255}$$

where f is the acceleration in units of the gravitational force. The factor $(1+f)$ would then be included in the substitutions following (247) so that, if we presume that some of the dynamical motions represented by dv/dt are dissipated in the form of heat, (249) becomes

$$E = f\Omega - \frac{\beta}{2}(1+f)\Omega + (1+f)\frac{8\pi G}{c^2}\int_0^R MPr\,dr + \\ +(f+\tfrac{1}{2})\frac{12\pi G^2}{c^2}\int_0^R \rho M^2\,dr + E(\text{dynamical}) \tag{256}$$

Then (250) becomes, with $\beta(1+f)/2 \ll f$,

$$\frac{E}{M_c^2} = \frac{3f}{2}\left(\frac{GM}{Rc^2}\right) + (5.1+9.2f)\left(\frac{GM}{Rc^2}\right)^2 + \frac{E(\text{dyn})}{Mc^2} \tag{257}$$

When the star is collapsing the acceleration is inward and f is negative; the collapse is free fall when $f=-1$ and hydrostatic equilibrium corresponds to $f=0$; the second term in (257) vanishes when $f=-0.55$. It follows that dynamical equilibrium corresponds to a state intermediate between hydrostatic equilibrium and free fall.

The introduction of a small amount of rotation also stabilises masses up to $10^8\ M\odot$, collapse proceeding along the axis of rotation. See, for example, W. A. Fowler 1966, *Astrophysical Journal 144*, 180.

It is not certain what will ultimately happen to the object; it seems likely that it will either explode or suffer fission. If the energy emitted by core collapse can be transferred efficiently to the envelope, the envelope may explode; thus as $2GM/Rc^2 \to 1$, $v(\text{collapse}) \to c$, then

$$t(\text{collapse}) \sim \frac{R}{c} \sim \frac{2GM}{c^3} \tag{258}$$

For $M=10^8\ M\odot$, this gives $t(\text{collapse}) \sim 10^3$ s. It is considered possible that some such event may account for the observations of objects known as 'quasars'.

The luminosity of the object prior to collapse can be calculated from the equation for a polytrope of index 3,

$$L = \frac{4\pi cGM(1-\beta)}{\overline{\kappa\eta}} \tag{259}$$

where $\overline{\kappa\eta} = (L_r/M_r)/(L/M) \sim$ unity. Since $\beta \to 0$, we have

$$\frac{L}{L\odot} \approx 10^4 \left(\frac{M}{M\odot}\right) \tag{260}$$

Thus a supermassive object is about ten thousand times more luminous per unit mass than an average dwarf star like the Sun, or an ordinary galaxy of stars.

7.6. *Nuclear Synthesis in Explosions.*

The central temperatures (figure 24) are $T_c \lesssim 2\times10^9$ K and typical densities (p. 97) are several g cm^{-3}. The nuclear reactions which operate in such conditions have already been dealt with in chapters 2 and 6; they follow the cycles

$$CN \to 3\alpha \to C^{12}(C^{12}, \gamma) \to Mg^{24} \to \text{Fe group}$$

the 3α cycle dominating in these conditions. The supermassive objects thus produce large amounts of helium and some heavier elements.

7.7. *Black Holes.*

It was pointed out in section 7.3 that when the gravitational potential energy of an object is comparable with its rest mass (equation 219), electromagnetic energy cannot escape from the object. A photon loses all its energy in doing work against the gravitational field before it can reach a great distance from the object. The light paths are curved to follow the surface of the object which, unable to emit radiation, is isolated from its environment. Mass and radiation can enter it, but none can escape; hence the term 'black hole'.

The confinement of radiation as well as mass, and its associated energy, within a spherical surface is somewhat analogous to a liquid drop where the surface energy is due to surface tension. This analogy has proved to be useful in interpreting the properties of black holes. Consider a black hole due to a total mass M, having angular velocity ω, angular momentum l, electromagnetic potential ϕ, charge Q, effective surface tension T and area A; then the energy relationship is

$$2TA+2\omega L+\phi Q=M \qquad (261)$$

It is instructive to compare this with the Virial, equation (46).

Caution is required in interpreting this equation. The area is that of a two-dimensional surface in space-time, representing the event horizon associated with the light paths discussed earlier in that it is the boundary from within which particles or photons can never escape; but with that qualification the analogy with the liquid drop is close. For example, as the angular momentum is increased the black hole flattens at the pole. If the angular momentum of a liquid drop is increased too much it becomes unstable and breaks up into two drops; it is possible that a black hole may similarly divide into two.

Although black holes cannot be observed directly they do affect the space surrounding them in a manner which may enable the existence of them to be inferred. For example, they deviate the paths of photons passing close to them so that objects seen beyond them may appear distorted, and the motions of visible objects may be similarly affected; but at the time of writing the existence of black holes has not been demonstrated by observation.